UNIVERSITY OF STRATHCLYDE

30125 00526680 3

Books are to be returned on or before
the last date below.

Influence and Removal of Organics in Drinking Water

Influence and Removal of Organics in Drinking Water

EDITED BY

J. Mallevialle
I. H. Suffet
U. S. Chan

LEWIS PUBLISHERS
Boca Raton   Ann Arbor   London   Tokyo

Library of Congress Cataloging-in-Publication Data

Influence and removal of organics in drinking water / edited by J.
  Mallevialle, I. H. Suffet, U. S. Chan.
     p.   cm.
  Includes bibliographical references and index.
  ISBN 0-87371-386-9
  1. Water--Purification--Organic compounds removal--Congresses.
  2. Drinking water--Purification--Congresses.   I. Mallevialle, Joël.
  II. Suffet, I. H.   III. Chan, U. S. (U. Sam)
  TD449.5.I53   1992
  621.1′62--dc20

92-7334
CIP

COPYRIGHT © 1992 by LEWIS PUBLISHERS
ALL RIGHTS RESERVED

  This book represents information obtained from authentic and highly regarded sources. Reprinted material is quoted with permission, and sources are indicated. A wide variety of references are listed. Every reasonable effort has been made to give reliable data and information, but the author and the publisher cannot assume responsibility for the validity of all materials or for the consequences of their use.

  Neither this book nor any part may be reproduced or transmitted in any form or by any means, electronic or mechanical, including photocopying, microfilming, and recording, or by any information storage and retrieval system, without permission in writing from the publisher.

  Direct all inquiries to CRC Press, Inc., 2000 Corporate Blvd., N.W., Boca Raton, Florida 33431

PRINTED IN MEXICO
1 2 3 4 5 6 7 8 9 0

Printed on acid-free paper

Preface

Frequently considered to be a symbol of purity, water has progressively become one of the most closely monitored elements in the food chain and is now subject to extremely stringent quality standards which have considerably increased the technical complexity of producing drinking water. This trend will continue and the production of water is rapidly becoming a high-tech industry.

One of the main difficulties lies in the presence of natural or anthropogenic organic matter in concentrations which, in natural water, can vary from the immeasurably small (nanograms per liter) in groundwater to some tens of milligrams per liter in surface waters. Unlike inorganic elements, it is not possible to prepare a comprehensive list of organic substances potentially present in the water. There could be several hundred thousand, or even several million, such substances, which makes them extremely difficult to identify.

The international scientific community has attached increasing importance to studying the nature of these organic compounds and their elimination since J.J. Rook discovered the formation of trihalomethanes (THMs) when water is chlorinated, in 1974. Whether directly (e.g., pesticides, polyaromatic hydrocarbons, etc.) or indirectly, through their reaction with chemicals required for disinfection (e.g., THMs, total organic halogens, etc.), they are at the origin of a number of short- or longer-term problems such as toxicity, tastes, and odors, or bacteria reviviscence.

This book contains the various papers presented as part of a workshop held in Macau between world experts in the elimination of organic matter from drinking water. It covers both fundamental aspects, such as adsorption or coagulation, and the results of experiments on, for example, ozone/granular activated carbon obtained from pilot plants.

A number of articles on new technologies, such as oxidizing coupling or membranes, give a glimpse of probable changes in the water treatment industry in the coming decade.

Joël Mallevialle
Mel Suffet

Joël Mallevialle graduated from the University of Toulouse, where he obtained a degree in chemical engineering in 1967. He then served as a research engineer at the Oceanographic Institute in Paris and later at the Organic Chemical Institute of the University of Zurich, where his research focused on the fate of biopterins in aquatic animals.

Since 1969 he has been at the Lyonnaise des Eaux/ Dumez company, where he first worked as a research engineer in the water field after he received his Ph.D. degree based on his studies on the ozonation of organic matters in water. In 1986 he became Deputy Manager of the Research Laboratory and in 1991, Director of the Center of International Research for Water and Environment (Lyonnaise des Eaux/Dumez).

Dr. Mallevialle has co-authored over 100 papers on the treatment of drinking water by ozonation, activated carbon, membranes, etc. and co-edited a book on tastes and odors in drinking water. He received the Chemviron Award in 1980 for the best European research in the water treatment field for 1978–1980. He is also a member of many international associations, including the American Water Works Association, the International Water Supply Association, and the American Chemical Society, and is presently a member of the Board of Directors of the International Ozone Association.

I.H. (Mel) Suffet is presently Professor, Environmental Science and Engineering Program, UCLA — School of Public Health. He most recently (1990) was the P.W. Purdom Professor of Environmental Chemistry at Drexel University. He received his Ph.D. from Rutgers University, his M.S. in chemistry from the University of Maryland, and his B.S. in chemistry from Brooklyn College. Among his numerous awards are the American Chemical Society's Zimmerman Award in Environmental Science, Drexel University Research Achievement Award, Pennsylvania Water Pollution Control Association Service Award, Instrument Society of America Service Award, and the Environmental Chemistry Division of the American Chemical Society Service Award.

Dr. Suffet has co-authored more than 100 research papers and monograph chapters on environmental and analytical chemistry. His research expertise in the field of environmental chemistry focuses on phase equilibria and transfer of organic chemicals. This expertise allows him to work on the analysis, fate, and treatment of organic chemicals and has led to his current studies on water treatment, including taste and odor problems and unit operations of sorption and chemical oxidation.

He has organized and chaired numerous technical society meetings. In addition, he has served on the Safe Drinking Water Committee of the National Academy of Sciences, for which he chaired the Subcommittee on Adsorption, and was a consultant to the Water Reuse Panel of the National Academy of Sciences.

Dr. Suffet co-edited, with J. Mallevialle, Identification and Treatment of *Tastes and Odors in Drinking Water* (1987) for the American Water Works Association Research Foundation; and seven other edited books on fate, treatment, and analysis of organic pollutants in water. He serves on the editorial boards of the journals *Chemosphere* and *CHEMTECH*. He completed a four-year term as treasurer of the American Chemical Society Division of Environmental Chemistry in 1988 and is presently the Secretary of the International Humic Substance Society.

 **U.S. Chan** graduated from the Mechanical Engineering Department of Northeast University of Technology in Shenyang, China. As an engineer, Mr. Chan worked in the brewery design field in the Light Industry Design Institute subordinated to The Ministry of Light Industry in China for 15 years. Mr. Chan then served as an associate scientist studying anaerobic digestion in the Guangzhou Institute of Energy Conversion subordinated to the Chinese Academy of Sciences for 10 years. He actively took part in the 2nd to 4th International Symposiums on Anaerobic Digestion as speaker and as a member of the organization committee.

Since 1982, Mr. Chan has worked in the Macao Water Supply Company. He is now the Chief Engineer of the company, in charge of all technical problems, including the research projects.

Acknowledgments

For a large project such as the one covered by this volume, many people have provided significant input. We cannot thank everyone in print because of space limitations, but we know and appreciate how important their contributions were to the project.

The workshop was co-sponsored by the Lyonnaise des Eaux/Dumez (France) and the Macao Water Supply Co. Ltd., which made it possible to attract "world class" practioners in the field to participate and to develop funds for publicity and publication.

We want to acknowledge for the Lyonnaise des Eaux/Dumez Company, Thierry Chambolle, Director of Research and Technological Development, and Michel Dupont, Director of our Far-East office, for their constant encouragement; Jean Pierre Duguet, Head of the Potable Water Department in our Research Laboratory, for his input into the organization of this conference; and Michel Hurtrez and Marie-Françoise Dumesnil for their graphic arts and editing assistance.

We would like to thank the Macao Water Supply Team, especially K.L. Chan, Managing Director, Yves Moyne, Executive Manager, and Paul Ledoigt, Technical Director, for their assistance and tireless efforts in making the symposium a success by providing an excellent conference facility and facilitating the meeting arrangements.

We would also like to acknowledge the expert reviewing and editing for some of the chapters by Mr. Lew Brenner of Drexel University's Environmental Studies Institute.

Finally, we wish to extend a special thanks to both of the organizations which have allowed this workshop.

Contents

1. Adsorption Thermodynamics at the Solid-Liquid Interface: The Influence of Surface Heterogeneity on the Removal of Some Surfactants and Organic Micropollutants from Aqueous Solutions, by *J.M. Cases* 1

2. Influence of the Concentration Change of Raw Water upon the Carbon Adsorption Isotherms of Total Organics and Micropollutants, by *A. Yuasa* 19

3. Using Powdered Activated Carbon: A Critical Review, by *I.N. Najm, V.L. Snoeyink, B.W. Lykins, Jr., and J.Q. Adams* 35

4. Development of a Method to Predict the Adsorption of Organic Chemicals on Activated Carbon, by *D.J.W. Blum and I.H. Suffet* 67

5. Competitive Adsorption of Several Organics and Heavy Metals on Activated Carbon in Water, by *Z.P. Jiang, Z.H. Yang, J.X. Yang, W.P. Zhu, and Z.S. Wang* 79

6. Coagulation-Flocculation of Minerals Using Al, Fe(III) Salts. What Kind of Flocs for What Separation Process? by *J.Y. Bottero* 97

7. Drinking Water Treatment by Hydrosoluble Polymers: Mechanisms of Coagulation and Efficiency, by *T.K. Wang, G. Durand, F. Lafuma, and R. Audebert* 115

8. The Role of Colloids and Dissolved Organics in Membrane and Depth Filtration, by *M.R. Wiesner* 125

9. Preliminary Analysis of the Organic Contents in Treated and Raw Water in Guangzhou City, by *L. Zhicai, C. Wanhua, and Z. Shaojia* 137

10. Treatability Evaluation by Simple and Rapid Method, by *N. Tambo and T. Kamei* 143

11. Automatic Control of the Coagulant Dose in Drinking Water Treatment, by *C. Hubele* 157

12. Influence of Preozonation on the Clarification Efficiency, by *Y. Richard* 173

13. Application of Ozone with Activated Carbon for Drinking Water Treatment in China, by *F. Jinchu and X. Jianhua* 187

14. Basic Concepts for the Choice and Design of Ozone Contactors, by *M. Roustan, E. Brodard, J.P. Duguet, and J. Mallevialle* 195

15. Organic Removal by Ozonation and GAC Adsorption, by *Y.-B. Feng, J.-S. Feng, and Y.-N. Wong* 207

16. A Pilot-Plant Study on Advanced Treatment of Potable Water, by *W. Dazhi and L. Bingjie* 219

17. Ozonation of Organic Compounds Causing Taste and Odor Problems, by *C. Anselme, J.P. Duguet, J. Mallevialle, and I.H. Suffet* 233

18. The Aqueous Reactions of Specific Organic Compounds with Ozone, by *N. Graham, C. Corless, G. Reynolds, R. Perry, and J. Haley* 251

19. New Advances in Oxidation Processes: The Use of the Ozone/Hydrogen Peroxide Combination for Micropollutant Removal in Drinking Water, by *J.P. Duguet, A. Bruchet, and J. Mallevialle* 265

20. A Fluidized Biofilter Bed Process as a Preliminary Treatment of a Contaminated Raw Water Source, by *Y. Zheng-Zhong and Z. Hua* 279

21. Removal of Synthetic and Natural Organic Compounds by Biological Filtration, by *B.E. Rittmann and J. Manem* 289

22. Membrane Filtration in Drinking Water Treatment: A Case Study, by *J. Mallevialle, J.L. Bersillon, C. Anselme, and P. Aptel* 299

23. Ultrafiltration of Lake Water: Optimization of TOC Removal and Flux, by *M.M. Clark* 311

24. Megatrends in Drinking Water Treatment Technologies: The Years Ahead 1990, by *F. Fiessinger* 325

Index 333

CHAPTER 1

Adsorption Thermodynamics at the Solid-Liquid Interface: The Influence of Surface Heterogeneity on the Removal of Some Surfactants and Organic Micropollutants from Aqueous Solutions

J. M. Cases

ABSTRACT

The removal of surfactants from aqueous solutions is a complicated phenomenon. Above the Krafft point, two extreme cases of surfactant adsorption on hydrophilic surfaces can be described: (1) at low relative concentration (high undersaturation relative to the critical micellar concentration), ionic surfactants with aliphatic chain ($n > 8$) which have a high affinity for the surface, were found to form monolayer aggregates (hemimicelles) whose sizes are limited by the surface heterogeneity, and (2) at high relative concentration (weak undersaturation), bilayer formation is possible if the surface is hydrophobic due to the first layer formation. For these two latter cases, the adsorption can be interpreted with the theory of two-dimensional condensation on a heterogeneous surface. In the case of adsorption on a hydrophilic surface, surfactants having a low affinity for the solid (i.e., nonionic surfactants or ionic surfactants) form micelles on surfaces which are morphologically similar to the aggregates that are formed in solution beyond the critical micellar concentration. These different possibilities for surface aggregation are due to the importance of the normal bond and the configurational freedom of hydrocarbon chains in the liquid crystal state. Below the Krafft point, the chain conformation (chains are extended in the transconformation and ordered) imposes the formation of bidimensionnal aggregates. In this case, precipitation can lead to multilayer formation. At low surface coverage, the removal from aqueous solutions of some organic micropollutants is also due to two-dimensional condensation (first-order change in the adsorbed layer due to lateral bonds).

INTRODUCTION

The physical adsorption of surfactants and organic micropollutants from aqueous solution is a phenomenon of major importance in many applications ranging from ore flotation to water treatment. This process is of considerable complexity and scientific interest. Modeling of the adsorption mechanism of surfactants has traditionally been based on the interpretation of the shape of the adsorption isotherms. Gaudin and Fuerstenau[1] described the role of lateral attractive bonds in the monolayer and the increase of surface coverage with formation of patches and proposed the hemimicelle theory.[2] More recently, Cases and co-workers[3-7] proposed the theory of two-dimensional condensation on heterogeneous surfaces that takes into account both potential energies and entropic terms in the adsorbed layer as well as the heterogeneity of the adsorbent surfaces. Two-dimensional condensation is possible for alkyl chains containing more than eight methylene groups. The size of the aggregates being a function of superficial heterogeneity, a single energetically homogeneous domain may be divided among one or more nonadjacent regions. The thermodynamic approach allows the state of condensation in the adsorbed layer (unknown) to be compared to a more easily characterized reference phase. Harwell et al.[8] proposed the pseudophase separation model in which the surface aggregation occurs in patches on heterogeneous surfaces, producing bilayered structures (admicelles). The apparent confusion in the literature regarding the mechanisms of surfactant adsorption and the interpretation of isotherm data reflects the complexity of the system studied. However very receivable direct methods of investigation — electron spin resonance, excitated resonance Raman spectroscopy, fluorescence decay spectroscopy (FDS) — have been used to study the adsorption of nonionic surfactants on silica[9,10] or the adsorption of an ionic surfactant on alumina.[11] The FDS method can be used to study the aggregates whose aggregation number is between 25 and 300 to 400. Micelles were found to form on the surface at a concentration slightly lower than the critical micellar concentration (C/CMC $> 0,1$).[10,12] In this paper, only the case of long chain surfactant is treated without taking into account either surface or bulk precipitation in order to avoid complicating the discussion.

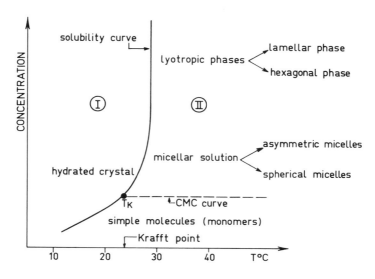

FIGURE 1. Schematic representation of the solubility of ionic surfactants vs temperature. Phase diagram with the Krafft point.

PHYSICOCHEMICAL PROPERTIES OF SURFACTANTS IN SOLUTION

An examination of the solubility curve of a surfactant as a function of temperature reveals two large domains (Figure 1). At temperatures below the Krafft point T_K (the characteristic transition temperature where the "chain melting" occurs) the solubility curve describes the saturation concentration of hydrated crystals in equilibrium with monomers. In the solid phase, the molecules are arranged in layers with saturated portions of the chain in the transconformation: the chains are extended and ordered corresponding to a cross-sectional area of 0.205 nm². At a temperature greater than the Krafft point (region II), the association of amphiphilic molecules in an aqueous medium typically occurs in a stepwise manner as the surfactant concentration increases. In a sufficiently dilute solution there are only surfactant monomers. As the concentration is increased, there is often association to form micellar aggregates which still exist in an isotropic solution. At still higher amphiphile concentrations, lyotropic liquid crystalline phases are formed.[13] For simple soaps, the phase formed first is the normal hexagonal phase consisting of cylindrical aggregates packed in a hexagonal array. At very high amphiphilic concentrations, a lamellar phase can appear. It also indicates that changes in the stability of different phase structures for a given amphiphile are not primarily caused by changes in local molecular properties, but rather in interactions associated with the aggregates themselves (bonds between counter ions and polar heads for instance). In these different phases the chains are liquid-like and highly disordered (liquid crystal state). The chain area in the condensed state,[6] corresponding to the lamellar phase, is only 33·Å². For these two extreme cases (liquid crystal or hydrated crystal), the calculation of the Gibbs energy change, Δg, of the transfer of one CH_2 from solution to condensed phase yields a value of 1.1 to 1.3 kT if the chain is flexible and a value greater than 1.4 kT if the chain is incorporated in a hydrated crystal or a precipitated phase.[6]

ADSORPTION OF SURFACTANTS AT THE SOLID-LIQUID INTERFACE

The adsorption of long chain ionic amphiphiles can be described by a model which takes into account the covering of the surface by surfactants as the result of a two-dimensional condensation on heterogeneous surfaces.[4,6,14] The thermodynamical treatment is based on the following considerations:

(1) Heterogeneous surfaces could be viewed as a combination of homogeneous ones. Each homogeneous domain i is characterized by its area S_i and the normal energy of adsorption $\phi^o_{a,i}$. The surface coverage, θ, can be expressed in the case of the two-dimensional condensation (a domain i is empty or full) by the relation:

$$\theta = \sum_1 S_i/S \tag{1}$$

where S is the total adsorbent surface.

(2) For a given homogeneous domain, the equilibrium equation of the adsorbed layer with the solution depends upon (a) the adsorption energy $\phi^o_{a,i}$ necessary to create a normal bond between the head of the surfactant and the site on the surface after removing water molecules; the bond can be of different types: electrostatic, hydrogen bonding . . . , (b) the sum, ω, of the lateral energies of one molecule located in a compact adsorbed layer (ω includes both the attractive hydrocarbon-hydrocarbon bonds and the repulsive head-head interaction), and (c) the sum for each molecule of some entropic terms such as internal and external vibrational entropies (T > or < T Krafft), conformational and rotational entropies (T > T Krafft) of the molecule in its adsorption site and the configurational entropy calculated from the Bragg-Williams approximation.

(3) If there are more than eight methylene groups in the aliphatic chain, the filling of an homogeneous domain results in a first order change. For a given concentration value $C^*_{e,n}$, the adsorbed layer swings abruptly from a dilute state ($\theta \approx 0$) to a condensed state ($\theta \approx 1.0$). Then the isotherm presents a small step (if S_i is sufficiently great) characterizing two-dimensional condensation on the domain i considered. Assuming that ω is a linear function of the number n of methylene groups in the apolar hydrocarbon chain, one obtains a relation similar to Traube's empirical law:

$$[kT \ln C^*_{e,n}]_\theta = -n\Delta g/2 + E_\theta \tag{2}$$

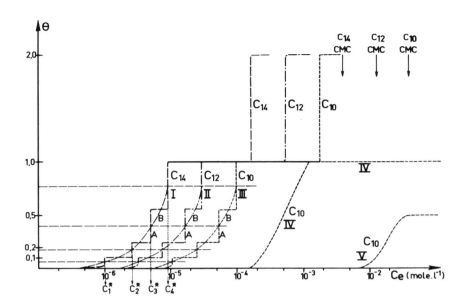

FIGURE 2. Schematic representation of adsorption isotherms on heterogeneous surfaces.

E_θ is a constant which includes $-\phi_{a,i}^\circ$. This relationship indicates that the greater the value of n, the lower the equilibrium concentration for which the characteristic step of the two-dimensional condensation occurs. For instance in Figure 2, curves I_A, II_A, and III_A, schematically represent the adsorption of a homologous series of surfactant with 10, 12, and 14 methylene groups on a heterogeneous surface formed with four homogeneous domains. Δg represents in fact the only parameter expressing (together with the area of the molecule when the monolayer is formed) the state of the adsorbed layer which can be measured from the experimental isotherms in the (θ, ln Ce) plane. Expression 2 is valid only if the amphiphilic aggregates (hemimicelles) contain sufficient molecules so that the properties of the aggregates approach those of macroscopic systems, i.e., the effects of the complex molecular interactions can be described with a few parameters due to overaging effects. If there is the same condensation state on the different homogeneous domains i, the slope of the straight line in the (ln Ce,n') plane must be the same ($-\Delta g/2$) whatever the θ value. The origins E_θ of the line will be different (Figure 3). When θ increases, the corresponding $\phi_{a,i}^\circ$ value of the different domains i decreases and E_θ increases.[4] This fact characterizes two-dimensional condensation on heterogeneous surfaces when the surface field does not influence the state of the layer. When the surface coverage is very low ($\theta \ll 1$), if Equation 2 is not verified, this indicates either that no aggregates are present on the surface, surfactants ions are then adsorbed as individual ions (dilute state), or aggregates are so small that the thermodynamics of infinite phase cannot be utilized.

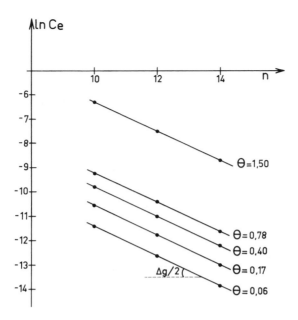

FIGURE 3. ln C_e profile as a function of the number of methylene groups in the aliphatic chain for different surface coverage values.

(4) It is also possible to know the state of the adsorbed layer by comparing its organization with that of a plane of a reference phase. These phases can be

- For $T <$ T Krafft, a reticular plane of the hydrated crystal (or precipitated phase)
- For $T >$ T Krafft, one layer of the lamellar lyotrope phase (liquid crystal state)

The comparison between the organization of the adsorbed layer and that of a plane of the reference phase can be reduced to a substraction between two equilibrium equations — one for the equilibrium of the adsorbed layer with the aqueous solution and the other for the equilibrium of the reference phase with its own aqueous solutions which saturation concentration is $C_{o,n}$.

$$kT \ln C_{e,n} - kT \ln C_{o,n} = kT \ln (C_{e,n}/C_{o,n}) = \Delta\mu \qquad (3)$$

In the case where $T > T_K$, for instance, it can be shown that the undersaturation $\Delta\mu^*$ corresponding to the step characteristic of two-dimensional condensation can also be expressed in terms of the critical micellar concentration.[4] A and B are constants for a homologous series of surfactants:

$$\Delta\mu'^* = kT \ln(C^*_{e,n}/CMC_n) = B - \varphi^o_{a,i} \qquad (4)$$

and

$$\Delta\mu^* = kT \ln C^*_{e,n}/C_{o,n} = A - \varphi^o_{a,i} \quad \text{if} \quad T < T_K \tag{4'}$$

If one assumes that the adsorbed layer and the reference phase are in the same state, it implies that the entropic terms, lateral energies, and coordination numbers are equal. This assumption is verified if:

1. The isotherms built with a homologous series of surfactants superimpose in the (θ, $\Delta\mu'$) plane, except for the lowest and the highest θ values. Generally, however, these differences at extreme θ values are not measurable when ω is great.
2. The values of Δg and the molecular area of the molecule in the closed sphere packing are the same for the adsorbed layer and the reference phase.
3. Equations 3 and 4 mean that the greater the $\phi^o_{a,i}$ values, the greater the undersaturation values (or the lower the relative concentrations) characteristic of two-dimensional condensation. The use of the (θ, $\Delta\mu'$) plane or (θ, $\Delta\mu$) plane when $T < T_K$ facilitates the evaluation of the normal bond, $\phi^o_{a,i}$ through the position of the isotherm on the undersaturation axis. It also indicates the state of the adsorbed layer in the case of superimposition.

On heterogenous surfaces, the shape of the isotherm according to Equations 1 and 4 depends on a double distribution: S_i area distribution on the θ axis and $\phi^o_{a,i}$ distribution on the $\Delta\mu'$ or $\ln C_{e,n}$ axis. Then surface heterogeneity can be studied from the shape of the experimental isotherms. The ionic surfactants which present a high value of $\phi^o_{a,i}$ should serve as tracers of surface heterogeneity, for instance to measure the area of broken edge faces on phyllosilicates[14] or the determination of the adsorption energy distribution.[4]

HOW DO SURFACTANTS ADSORB ON A HYDROPHILIC SURFACE?

The Isotherms are Preformed at a Temperature Higher than the Krafft Point

The use of a surfactant with 10 methylene groups as adsorbate on different hydrophilic solids, α, β, γ, can lead to the isotherms schematically presented in Figure 2 (curves III$_B$, IV, V). The isotherms are here smoothed because the solid surfaces, taken as an example, are very heterogeneous and formed of a great number of homogeneous domains. Each point of the isotherms represents a particular condensed homogeneous domain. Considering the value of the CMC of the adsorbate and Equations 3 and 4, it can be noted that the different domains are filled as $\phi^o_{a,i}$ decreases. If the surface field does not influence the organization

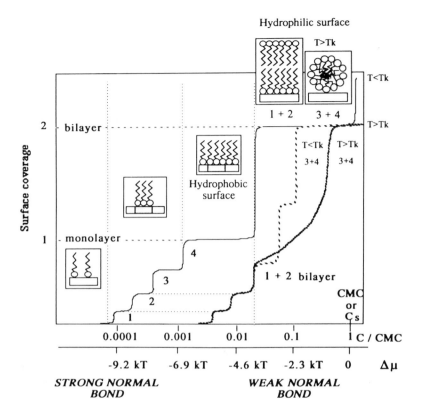

FIGURE 4. Schematic representation of different types of adsorption isotherms on heterogeneous hydrophilic surfaces.

of the adsorbed layer, the state of the layer (here the liquid-crystal state) is the same whatever domain i is filled as justified on Figure 3 (Δg is nonsurface coverage-dependent). Depending of the value of $\phi_{a,i}^{o}$, the different affinity of the surfactant for solids should be classified as follows:

1. Strong affinity for solid α (Figure 2, curve III_B and Figure 4) Ce/CMC < 0.01
2. Medium affinity for solid β (Figure 2, curve IV) 0.01 < Ce/CMC < 0.1
3. Weak affinity for solid γ (Figure 2, curve V)

In the last case, the surface cannot be covered with a monolayer at the CMC.

Cases 1 and 2 lead to the filling of the first adsorbed layer by formation of patches and it is then possible to apply all the concepts presented here. The association of adsorbed ions to form two-dimensional aggregates, that can be called hemimicelles for convenience, are:

- The importance of the normal bond $\phi_{a,i}^{\circ}$ that forces the surfactant to adsorb on the solid with the hydrophilic group in the amphiphilic molecule fixed at the water-solid interface and the hydrocarbon chains projected outward. Thus, the surface becomes hydrophobic as the surface coverage increases up to the monolayer capacity.
- The importance of lateral attractive bonds which leads to two-dimensional condensation and formation of patches.
- The surface heterogeneity which controls the size of the aggregates. At the beginning of the isotherms, the most energetic domains are filled; these domains are necessarily of weak extension (low area) and the number of aggregation could be limited to values lower than 1000. On the opposite, near the monolayer capacity, the domains are of lower energy, i.e., large area is filled. In this case, the aggregation number is very high. A point of the isotherm of 0.01 θ in extension corresponds to a homogeneous domain filled with 3×10^{16} molecules. For this reason the thermodynamics of infinite phases can be used to explain the filling of the first layer, except for the lowest surface coverages: in this case, Equation 2 is not valid.

The formation of the bilayer is always characterized[4] by a vertical step of the isotherm because the second layer occurs on a surface now energetically homogeneous, due to the first layer formation (Figures 2 and 4). The surface is now hydrophilic again because of the lyophilic groups accumulated in the liquid-like inner part of the bilayer. The hydrophilic groups are directed towards water.

When the affinity of the polar head of the surfactant is medium, the plateau corresponding to the monolayer capacity can be reduced (Figure 2, curve IV) or can even disappear if the undersaturation corresponding to the bilayer formation (and the formation of the tail-tail bonds) is reached before monolayer completion. For low affinity systems (weak normal bond) (Figure 2, curve V or Figure 4), two-dimensional condensation cannot occur. Recently, fluorescence and electron spin resonance spectroscopic probes have been used to study the solid water interfacial layer formed by dodecylsulfate adsorbed on alumina. The authors think that these methods are sensitive techniques to monitor the formation of hemimicelles.[11] Their experiments suggest that at low surface coverage, hemimicelles form at a relative constant size (approximately 100 SDS hemimicelles) and at high surface coverage; addition of more SDS molecules leads to an increase in the size of existing hemimicelles (up to 356 SDS per hemimicelle) rather than in the number of such species. A few years before Levitz and co-workers[9,10,12] extended the use of the fluorescence decay method developed by others for the determination of micellar aggregation numbers to the study of the adsorbed phase formed by nonionic surfactants (two monodisperse and four polydisperse of triton family polymethylene alkyl phenols) at the silica-aqueous solutions interface.

This method is based on the exclusive solubilization of a strong hydrophobic fluorescent probe molecule (pyrene) within the hydrophobic regions of the condensed molecular assemblies of the adsorbed layer. The simultaneous analysis of the time laws for the pyrene showed that from $\theta = 0.07$ to 0.8, the adsorbed layer has a fragmented structure, i.e., molecular aggregates described as not perfectly spherical micelles, but most probably oblate ellipsoids, whereas beyond $\theta = 0.8$ it is better described as an infinite or continuous medium in which large diffusion paths exist for the probe molecule. This region can correspond to a bilayer by melting of the micelles. The parallel determination of the average aggregation number and aggregate number density in the fragmentation phase domain shows three distinct regimes: (1) a growing regime from $\theta = 0$ to 0.17, where the size of the surface aggregates increases at constant number density; (2) a self-repeating regime from 0.17 to $\theta \sim 0.5$, where the number density increases while the size of the aggregates remains constant. The surface aggregates in this regime are undistinguishable from the regular micelles formed by TX 100 in aqueous solution (aggregation number $= 150$); and (3) again a growing regime to a point $\theta = 0.8$, where the adsorbed phase can be described as an assembly of surface micelles ($N = 200$) in close packing. The aggregation number is strongly dependent on the length of the polar chain. It decreases from 200 to 40 as the length increases from 9.5 to 40 OE units.

Taking into account the opposite results found for the domains where the surfactant presents a weak affinity for the solid surface, by Chandar and by Levitz and co-workers using the same method of investigation (FDS), a first question can be

> Are the results obtained by Levitz and co-workers a fundamental characteristic of the adsorption of nonionic surfactants on an hydrophilic surface? The adsorption process which occurs mainly below the critical micellar concentration is strikingly similar to the micellization process which takes place in water above the CMC. Below the CMC, the surface of silica acts as a precursor of the micellization process. The formation of micelles on the surface was corroborated by Desnoyel[15] using microcalorimetric measurements: this phenomenon corresponds to an endothermic signal analogous to that observed for micelle formation of nonionic surfactants above the CMC.

The second question could be

> Why are hemimicelles and not micelles described by Somasundaran and co-workers? To interpret their results these authors take into account (1) the difference viscosity obtained for surface aggregate: 100 cp for SDS hemimicelles compared to about 10 cp for SDS micelles. For motions of the solubilized probe, hemimicelles appear more rigid at the microscopic level[16] and (2) the difference obtained with excited state Raman spectroscopy by observing the Raman spectrum of tris(2.2′ bipyridyl)

ruthenium(II) incorporated in the solid-liquid interface under in situ equilibrium conditions,[17] the 1286 cm^{-1} peak observed for SDS micelles shift to 1281 cm^{-1} in hemimicelles. One can answer to the second question that such small differences could be obtained with micelles slightly perturbed due to the solid surface field. This would explain the weak aggregation number observed (max 50 to 250). In this range, if hemimicelles exist corresponding to an hydrophobic surface, their aggregation number would be high due to the solid surface formed of domains of weak energy, i.e., large extension. The answer to the first question could be the following: the origin of the difference in type of aggregates is to be found in the value of the normal adsorbate-adsorbent bond and not in the type of surfactant (ionic or nonionic). For strong or medium values, the aggregate has a two-dimensional extension because the surfactant polar head affinity is greater for the surface than for water: this bond forces the surfactant molecule to present its hydrophobic chain towards aqueous solution. On the other hand, a weak interaction between the polar point of the surfactant and the solid surface allows a rebuilding of a micellar-like aggregate below, but near the CMC, without a deep perturbation due to the solid surface field in such a way as to protect the aliphatic chains from water. In this last case, the affinity of the surfactant for the surface is close to the water one.

The Isotherms are Preformed at a Temperature Below the Krafft Point

As the chain is always extended in a condensed state, two-dimensional aggregates only are formed on the surface and the theory of the two-dimensional condensation on heterogeneous surface can be applied. An interesting case is described and presented in Figure 4. If the monolayer capacity is not reached for the value of undersaturation corresponding to the formation of bilayer (the normal bond $\phi_{a,2}^{o}$ corresponds to tail-tail bonds), the residual empty domain (here 3 and 4) could be filled directly with bilayer.[8] But this hypothesis is in fact a natural consequence of the two-dimensional condensation on heterogeneous surfaces.

APPLICATION TO THE REMOVAL OF SOME ORGANIC REAGENTS IN WATER TREATMENT

The Removal of Long Chain Sodium Soaps During Flocculation with Aluminum Hydroxide Gels

The adsorption of sodium decanoate and sodium dodecanoate on aluminum hydroxide gels at various pH values was studied by the difference of solution

Table 1. Experimental Values of Δg and Ordinate Values at Origin E

		pH 6.5		pH 7.5	
r	θ	$-\Delta g \times kT$	$E \times kT$	$-\Delta g \times kT$	$E \times kT$
0	0.1	1.408	−2.283	1.427	−2.095
	0.2	1.442	−1.931	1.445	−1.677
	0.35	1.415	−1.812	1.462	−1.219
	0.50	1.422	−1.586	1.478	−0.818
	0.65	1.459	−1.335	1.447	−0.782
	0.80	1.485	−1.189	1.453	−0.628
	Mean	1.438		1.452	
	Value	±0.027		±0.015	
2.0	0.1	1.137	−4.72	1.486	−2.379
	0.2	1.131	−4.59	1.455	−2.256
	0.35	1.111	−4.44	1.430	−1.939
	0.50	1.127	−4.08	1.431	−1.745
	0.65	1.115	−3.85	1.465	−1.552
	0.80	1.119	−3.62	1.484	−1.432
	Mean	1.141		1.458	
	Value	±0.025		±0.022	
2.5	0.1	1.477	−2.646	1.479	−1.965
	0.2	1.453	−2.600	1.501	−1.663
	0.35	1.433	−2.497	1.471	−1.596
	0.50	1.437	−2.271	1.481	−1.382
	0.65	1.479	−1.965	1.531	−1.010
	0.80	1.479	−1.965	1.632	−0.403
	Mean	1.459		1.517	
	Value	±0.019		±0.055	

concentrations before and after adsorption and the electrophoretic mobility method.[18] The gels were produced by adequate dilutions of coagulating solutions obtained by neutralizing 0.1 M aluminum chloride solutions with sodium hydroxide. Coagulating solutions were characterized by the neutralization ratio r = (NaOH)/(Al$_T$) with r = 0; r = 2.0; r = 2.5.[19] From the isotherms, it is possible to calculate Δg, the free energy change of transfer of one CH_2 group from the adsorbed layer to solution using Equation 2. The different results are presented in Table 1. The values of Δg obtained, $1.40 < \Delta g < 1.52$, allow us to keep the value 20.5 Å^2 for the molecular area of adsorbate in the condensed state, i.e., the hydrated crystal state. In this condition, plateaus of experimental isotherms correspond to the monolayer capacity that confirms the choice of the state of the adsorbed layer (Figure 5). Then, it is possible to calculate the specific surface area of the adsorbent (Table 2). The affinity of the adsorbate (A) for the adsorbent decreases as follows:

$$A_{r=2} > A_{r=2.5} > A_{r=0}$$

The normal bond, $\phi^o_{a,i}$, is entirely of electrostatic origin because the monolayer capacity always corresponds to the point of zeta reversal (PZR) (Figures 5 and 6). The bilayer formation corresponds to a vertical step (first-order change, the second layer swings between a dilute state and a condensed state) characteristic

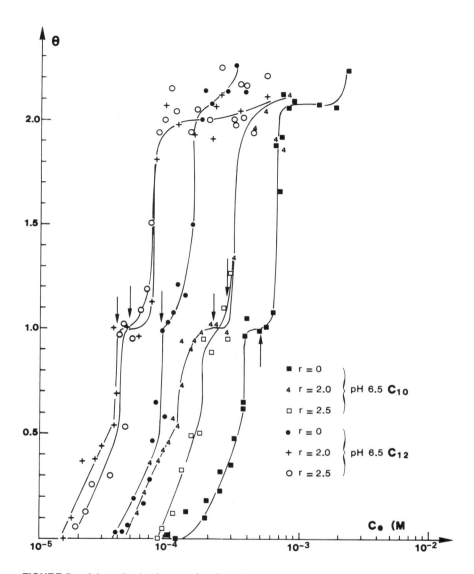

FIGURE 5. Adsorption isotherms of sodium decanoate and dodecanoate in the (θ, log C_e) plane at pH 6.5. Arrows indicate PZR position.

Table 2. Specific Surface Areas ($m^2\ g^{-1}$) of Gels Calculated from Adsorption Isotherms

r	Sodium decanoate			Sodium dodecanoate			Mean values		
	0	2.0	2.5	0	2.0	2.5	0	2.0	2.5
pH 6.5	1340	786	576	1064	726	610	1202	753	593
pH 7.5	689	640	582	582	576	501	636	608	542

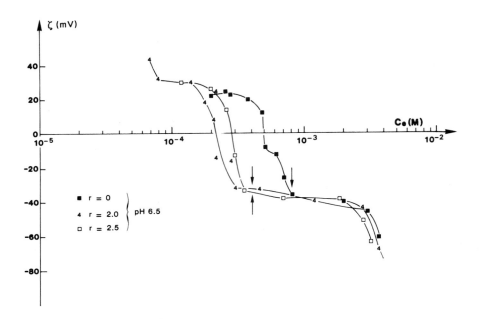

FIGURE 6. Variation of electrokinetic potential of flocs vs equilibrium concentration of sodium decanoate at pH 6.5. Arrows indicate bilayer capacity.

of two-dimensional condensation on a homogeneous surface. The surface is now homogeneous due to the first adsorbed layer formation. After the bilayer formation, zeta potential is constant (Figure 6) up to a sharp decrease due to the precipitation of aluminum soaps.

The Removal of Diethylphtalate and Dibutylphtalate

Coagulating solution was characterized by a neutralization ratio $r = (OH)/(Al_T) = 2.5$ and a total aluminum content of $0.15\ M$.[20] The tests were conducted on water from the western Paris suburb distribution system (Le Pecq - 78230 - France). This water which did not contain any mineral colloids, was artificially polluted with micropollutants of a desired concentration. The physicochemical characteristics of the water were as follows: pH 8.0, resistivity = 1568 at 20°C, TH = 32.2°Fr, TAC = 22°2 Fr, $Cl^- = 30$ mg/L,$^{-1}$ $SO_4 = 85$ mg/L^{-1}. The determinations of the equilibrium concentrations were made by gas-liquid chromatography.[21] Isotherms present in Figure 7 show that even at low surface coverage (here $0.01 < \theta < 0.04$), one observes the step characteristic of the two-dimensional condensation on high energetic homogeneous domain and not a Langmuir isotherm or in this low surface coverage domain a Henry's law. The first-order change is imposed by the importance of lateral bonds between CH_2,

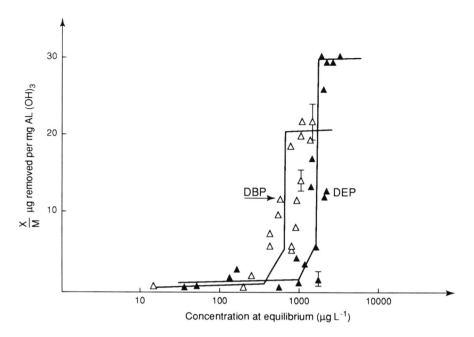

FIGURE 7. Adsorption isotherms of phtalates on aluminum hydroxide flocs (r = 2.5; pH = 8.0).

CH_3 groups, and benzene rings. The use of Equation 2 allows the calculation of Δg. The value, 0.98 kT, obtained seems to indicate probably a liquid crystal state of the adsorbed layer.

CONCLUSIONS

In water treatment, often the isotherms are given in the plot of (Q_a) logarithm of the adsorbed amount against the logarithm of the equilibrium concentration. If the isotherms are linear, they are characterized at first approximation by a Freundlich type equation of the form $Q_a = k\, C_e^{1/n}$, k and n being constant characteristics of the systems studied. To establish a Freundlich isotherm, it is necessary to assume the surface heterogeneity. This use is convenient to characterize the affinity of the adsorbate for the adsorbent, particularly with the n value, but completely screens the real mechanisms involved in adsorption; for instance, the possible influence of lateral bond and surface heterogeneity, as it has been demonstrated in this paper. The best knowledge of the phenomena involved in adsorption should help to predict and improve the efficiency of new coagulant species to fight against and control pollution.

REFERENCES

1. Gaudin, A. M. and Fuerstenau, D. W. "Quartz Flotation with Cationic Collectors," *Trans. AIME* 202:958-962 (1955).
2. Fuerstenau, D. W. and Raghavan, S. "Some Aspects of the Thermodynamics of Flotation," in *Flotation*, Fuerstenau, M. C., Ed. (New York: AIME, 1976), pp. 21-65.
3. Cases, J. M. and Mutaftschiev, B. "Adsorption et Condensation des Chlorhydrates d'Alkylamine à l'Interface Solide-Liquide," *Surface Sci.* 9:57-72 (1968).
4. Cases, J. M. "Adsorption des Tensio-Actifs à l'Interface Solide-Liquide: Thermodynamique et Influence de l'Hétérogénéité des Adsorbants," *Bull. Minér.* 102:684-707 (1979).
5. Cases, J. M., Canet, D., Doerler, N., and Poirier, J. E. "Adsorption des Tensio-Actifs Ioniques à Chaîne Alkyle à l'Interface Solides-Solutions Aqueuses," in *Adsorption at the Gas-Solid and Liquid-Solid Interface*, Rouquerol, J. and Sing, K. S. W., Eds. (Amsterdam: Elsevier/North-Holland, 1982), pp. 21-53.
6. Cases, J. M., Poirier, J. E., and Canet, D. "Adsorption à l' Interface Solide-Solution Aqueuse des Tensio-Actifs Ioniques: Les Systèmes à Forte Liaison Normale Adsorbat-Adsorbant et Surface Hétérogène," in *Solid-Liquid Interactions in Porous Media*, (Paris: Technip, 1985), pp. 335-370.
7. Cases, J. M., Levitz, P., Poirier, J. E., and Van Damme, H. "Adsorption of Ionic and Nonionic Surfactants on Mineral Solids from Aqueous Solutions," in *Advances in Mineral Processing*, Somasundaran, P., Ed. (Littleton: SME Publishers, 1986) pp. 171-188.
8. Harwell, J. H., Schechter, R. E., and Wade, W. H. "Recent Developments in the Theory of Surfactants Adsorption on Oxides," in *Solid-Liquid Interactions in Porous Media*, (Paris: Technip, 1985), pp. 371-410.
9. Levitz, P., El Miri, A. L., Keravis, D., and Van Damme, V. H. "Adsorption of Nonionic Surfactants at the Solid Solution Interface and Micellization: A Comparative Fluorescence Decay Study," *J. Colloid Interface Sci.* 99:484-492 (1984).
10. Levitz, P., Van Damme, H., and Keravis, D. "Fluorescence Decay Study of the Adsorption of Nonionic Surfactants at the Solid-Liquid Interface. I. Structure of the Adsorption Layer on a Hydrophilic Solid," *J. Phys. Chem.* 88:2228-2235 (1984).
11. Chandar, P., Somasundaran, P., and Turco, N. J. "Fluorescence Probe Studies on the Structure of Adsorbed Layer of Dodecylsulfate at the Alumina-Water Interface," *J. Colloid Interface Sci.* 117:31-46 (1987).
12. Levitz, P. and Van Damme, H. "Fluorescence Decay Study of the Adsorption of Nonionic Surfactants at the Solid-Liquid Interface. II. Influence of Polar Chain Length," *J. Phys. Chem.* 90:1302-1310 (1986).
13. Marchal, J. P., Canet, D., Néry, H., Robin-Lherbier, B., and Cases, J. M. "Nuclear Magnetic Resonance Study of Alkylammonium Chlorides. III. Micelle Characterization and Dynamic Properties Determined by Proton and Carbon-13 Longitudinal Relaxation," *J. Colloid Interface Sci.* 82:349-359 (1984).
14. Cases, J. M., Cunin, P., Grillet, Y., Poinsignon, C., and Yvon, J. "Methods of Analyzing Morphology of Kaolintes: Relation Between Crystallographic and Morphological Properties," *Clay Miner.* 21:55-69 (1986).

15. Desnoyel, R. "Etude Thermodynamique de l'Adsorption de Molécules Tensio-Actives à l'Interface Oxyde Minéral-Solution Aqueuse," *Thèse d'Etat,* Marseille I, 165 pages.
16. Somasundaran, P., Chandar, P., Turco, N. J., and Waterman, K. C. "Microstructure of Surfactant Adsorbed Layer at the Solid-Liquid Interface: Fluorescence and E.S.R. Investigation," in Forssberg, *XVIth Int. Min. Proc. Congress* (Amsterdam: Elsevier/North-Holland, 1988), pp. 77-784.
17. Somasundaran P., Kunjappu, J. T., Kumar, C. V., Turco, N. J., and Bacton, J. K. "Excitated State Resonance Raman Spectroscopy as a Probe of Alumina-Sodium Dodecylsulfate Hemimicelles," *Langmuir* 5:215-218 (1989).
18. Rakotonarivo, E., Bottero, J. Y., Cases, J. M., and Leprince, A. "Study of the Adsorption of Long Chain Sodium Soaps from Aqueous Solutions on Aluminum Hydroxide Gels," *Colloids Surfaces* 16:153-173 (1985).
19. Bottero, J. Y., Cases, J. M., Fiessinger, F., and Poirier, J. E. "Studies of Hydrolyzed Aluminum Chloride Solutions. I. Nature of Aluminum Species and Composition of Aqueous Solutions," *J. Phys. Chem.* 84:2933-2939 (1980).
20. Thébaut, P., Cases, J. M., and Fiessinger, F., "Mechanism Underlying the Removal of Organic Micropollutants during Flocculation by an Aluminum or Iron Salt," *Water Res.* 15:183-189 (1981).
21. Thébaut, P. "Elimination des Matières Organiques et des Micropolluants Contenus dans les Eaux de Surface: Rôle et Efficacité des Coagulants Minéraux," *Thèse de Docteur-Ingénieur INPL,* Nancy, 261 pages.

CHAPTER 2

Influence of the Concentration Change of Raw Water upon the Carbon Adsorption Isotherms of Total Organics and Micropollutants

A. Yuasa

ABSTRACT

IAS theory was applied to the multisolute competitive adsorption on activated carbon. Numerical simulations were executed for an imaginary raw water containing solutes, the single-solute isotherms of which are given by Freundlich equations. The result showed that the batch adsorption isotherm of total solutes changes when the raw water is diluted or concentrated; however, the modified isotherm does not change. This modification method was successfully applied to the experimental batch adsorption isotherm of humic acid measured by UV-absorbance.

Numerical simulations also showed that the spreading pressure which specifies the adsorbed phase can be correlated with the total solid-phase concentration. Accordingly the ratio of the solid-phase concentration to the liquid-phase concentration of each solute can be correlated with the total solid-phase concentration. The latter correlation was successfully applied to the experimental batch adsorption equilibria of p-aminobenzoic acid in the Seine River water.

PURPOSE OF STUDY

IAS (ideal adsorbed solution) theory has been recently applied to multisolute adsorption equilibria.[1-12] It is not practical, however, to apply IAS theory strictly to actual water because it is almost impossible to qualify all compounds and their adsorbabilities present in actual water and wastewater.[12] Practical interests are mainly focused on the behavior of total organics represented by surrogate parameters such as TOC, COD, UV-absorbance, etc. and that of unfavorable micropollutants of special concern such as toxic or carcinogenic compounds. This study aims to give a simple model which predicts the influence of dilution or concentration of raw water on the adsorption isotherm of total organics. The second purpose is to propose a model to describe the influence of background organics on the adsorption equilibrium of a certain component. For these purposes, numerical simulations based on IAS theory were executed on an imaginary mixture. Batch adsorption isotherm tests of humic acid in pure water and p-aminobenzoic acid in the Seine River water were carried out to verify the models.

IAS THEORY

Equations of IAS theory are given as follows:[1,2]

$$C_i = (q_i/q_t)C_i^* \qquad (i = 1,2,\ldots,N) \qquad (1)$$

$$\Sigma(q_j/q_j^*) = 1 \qquad (j = 1,2,\ldots,N) \qquad (2)$$

$$\pi A/RT = func(q_i^*, C_i^*) \qquad (3)$$

$$q_i^* = f_i^*(C_i^*) \qquad \text{(single-solute isotherm)} \qquad (4)$$

$$q_t = \Sigma q_j \qquad (5)$$

C_i and q_i are the liquid-phase and the solid-phase concentrations of solute i, respectively, and π is the spreading pressure at an equilibrium state in a multisolute system. C_i^* and q_i^* are the concentrations which give the same value of π in each single-solute system.

If the single-solute isotherms are described by the Freundlich equations as given by:

$$q_i^* = K_i(C_i^*)^{1/n_i} \qquad (i = 1,2,\ldots,N) \qquad (6)$$

IAS theory Equations 1 through 5 give:[11,12]

$$C_i = (q_i/q_t)[(\pi A/RT)/(K_i n_i)]^{n_i} \qquad (7)$$

$$\pi A/RT = n_i q_i^* = \Sigma(n_j q_j) \qquad (8)$$

Table 1. Composition of the Imaginary Raw Water YA100

i	K_i (μmol/g)	n_i	C_{io} (μmol/L)	q_{io}^* (μmol/g)
1	5	0.5	5	11.2
2	10	0.3	5	16.2
3	15	0.5	5	33.5
4	20	0.4	5	38.1
5	30	0.2	5	41.4
6	40	0.4	5	76.1
7	50	0.3	5	81.0
8	60	0.2	5	82.8
9	80	0.4	5	152
10	100	0.3	5	162
11	150	0.5	5	335
12	200	0.4	5	380
13	300	0.2	5	414
14	400	0.4	5	761
15	500	0.3	5	810
16	600	0.2	5	828
17	800	0.4	5	1523
18	1000	0.3	5	1621
19	1500	0.5	5	3354
20	2000	0.3	5	3241

i: Solute number.
K_i and n_i: Freundlich constants of solute i in the single-solute system.
C_{io}: Initial concentration of solute i.
q_{io}^*: Solid-phase concentration of solute i in equilibrium with C_{io} in the single-solute system.

The equilibrium state in batch adsorption is determined by solving Equation 6 and the following mass balance equation for N solutes simultaneously.

$$q_i = (C_{io} - C_i) V/M, \quad M \neq 0 \quad (i = 1,2,\ldots,N) \quad (9)$$

The solution q_{io} for $V/M = \infty$ is given by substituting $C_i = C_{io}$ to Equation 7 which represents the state of complete exhaustion of a column adsorber.

EFFECT OF DILUTION OR CONCENTRATION OF RAW WATER ON THE ADSORPTION ISOTHERM OF TOTAL SOLUTES

Simulation for an Imaginary Raw Water

The composition of the imaginary raw water named YA100 is shown in Table 1. The Freundlich constants for each component are given. The value of K_i increases as the component number i increases. The value of n_i is set within the range of 0.2 to 0.5 as encountered in most organic compounds. The initial concentration (C_{io}) is 5 μmol/L for each solute and the total initial concentration

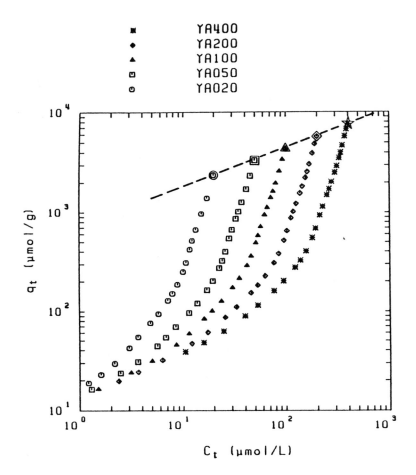

FIGURE 1. Influence of dilution and concentration of raw water on the isotherm of total solutes.

(C_{to}) is 100 μmol/L. Other raw water samples named YA020, YA050, YA200, and YA400 were prepared by diluting and concentrating YA100 so that C_{to} becomes 20, 50, 200, and 400 μmol/L, respectively. Simulations of batch adsorption equilibria for these imaginary waters were executed at carbon doses ranging 0 to 10 g/L.

The results are shown in Figure 1. The batch isotherm of total solutes for each raw water is given by plots of the same mark and the value of q_{to} ($= \Sigma q_{io}$) is given by a double plot such as ⊚. The broken line shows the isotherm obtainable from column tests in which diluted or concentrated raw water is used as the influent, and this isotherm can be represented by a Freundlich equation. At the points on the column isotherm, very strong adsorbable components predominate on the solid-phase. The distribution of mole fraction in the solid-phase

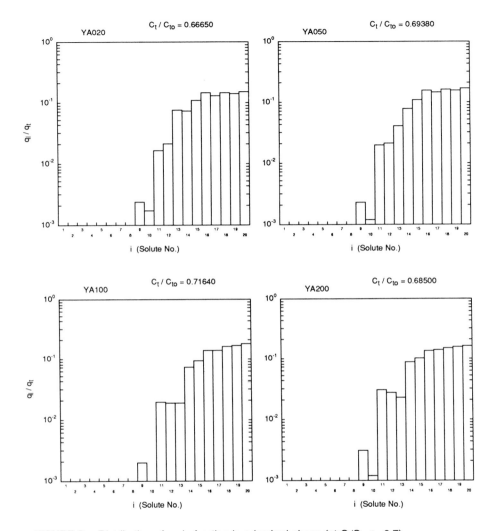

FIGURE 2. Distribution of mole fraction in adsorbed phase (at $C_t/C_{to} \fallingdotseq 0.7$).

is very similar for every other raw water when the ratio C_t/C_{to} is similar, as shown in Figure 2. The shapes of the batch isotherms in Figure 1 are very similar to each other and therefore the modified isotherms (q_t/q_{to} vs C_t/C_{to}) converge to one line as shown in Figure 3. The convergency was not a special case of this imaginary water YA100 and was observed for other mixtures of 10 to 40 components defined in different ways.

This model (the convergence of the modified isotherm) is, however, very limited when applied to actual water, because the value of q_t is unknown on a molar basis necessary for strict discussions based on IAS theory and is measurable only on the basis of surrogate parameter. In order to know the possibility of this

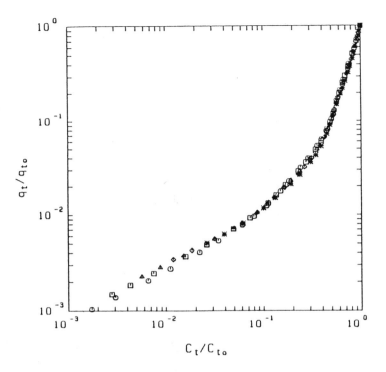

FIGURE 3. Modified isotherms.

model on the latter basis, the isotherms in Figure 1 were converted to those on the basis of UV-absorbance, assuming the molar absorbance ϵ_i for each solute randomly:

$$C_T = \Sigma(\epsilon_i C_i) \qquad (i = 1,2,\ldots,N) \qquad (10)$$

$$q_T = \Sigma(\epsilon_i q_i) \qquad (11)$$

Many sets of ϵ_i values were examined, and the modified isotherms based on UV (q_T/q_{TO} vs C_T/C_{TO}) converged to almost one line when the set of ϵ_i values was so defined that the ratio of the maximum to the minimum ($\epsilon_{max}/\epsilon_{min}$) is in the range of 5 to 100. In addition, the values of q_{to} (or q_{TO}) cannot be obtained by batch tests and are measurable only inaccurately by column tests because of the long-tailed breakthrough for actual water. If the column isotherm can be represented by the Freundlich equation as given by:

$$q_{TO} = KC_{TO}^{1/n} \qquad (12)$$

the plots of $q_T/C_{TO}^{1/n}$ vs C_T/C_{TO} give also a converged isotherm when the value of $1/n$ is properly estimated.

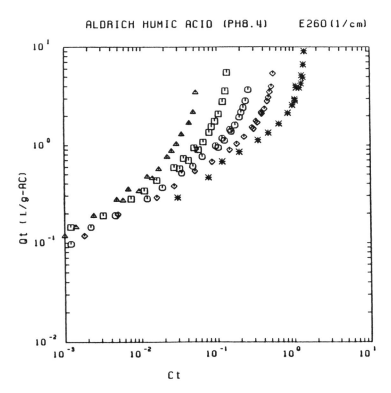

FIGURE 4. Influence of the initial concentration of humic acid on the isotherm measured by UV-absorbance (at 260 nm, 1-cm cell).

Batch Isotherm of Commercial Humic Acid

Experiments of batch adsorption of humic acid on activated carbon were carried out. Humic acid (Aldrich Chemical Company) was dissolved in pure water (GAC-treated, ion-exchanged, and distilled water) and was buffered to pH 8.4 by 0.01 M $NaHCO_3$. The initial concentration of humic acid is 2, 5, 10, 20, and 50 mg/L, and the initial UV-absorbance (at 260 nm, 1-cm cell) was 0.060, 0.151, 0.306, 0.595, and 1.478, respectively. Filtrasorb 400 (Calgon Co. Ltd.) was used after pulverized to less than 74 μm. The carbon dose ranged from 2 mg/L to 10 g/L. Each bottle (500-mL Erlenmeyer flask) contained 200 mL of the solution and was shaken for 8 days in a room maintained at 20°C.

Experimental results are shown in Figure 4. Assuming the value of 1/n (= 0.05, 0.1, 0.15, . . . , 1.0), the modified isotherms [$q_T/C_{TO}^{1/n}$ vs C_T/C_{TO}]

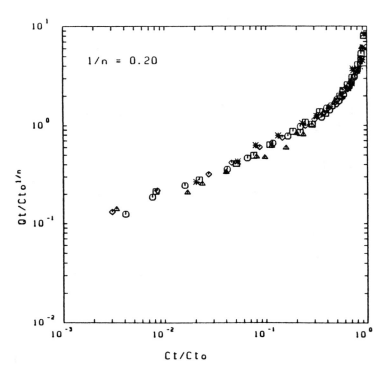

FIGURE 5. Modified isotherm of humic acid.

based on UV-absorbance were plotted. The best convergence of the isotherm was obtained when 1/n is 0.20 or 25 as shown in Figure 5. The convergence is also fairly good when 1/n is between 0.15 and 0.30. Therefore, a rough estimation of the value of 1/n may be sufficient for actual water.

INFLUENCE OF BACKGROUND ORGANICS UPON THE ADSORPTION EQUILIBRIUM OF A CERTAIN COMPONENT

Because the spreading pressure is a one-valued function of C_i^* and vice versa, Equation 1 is written in a general form as follows:

$$q_i/C_i = q_t \, func_i(\pi) \qquad (13)$$

This shows that the distribution ratio of each solute in the solid and the liquid phases is governed by the total solid-phase concentration and the spreading pressure. If the spreading pressure is correlated with the total solid-phase concentration, Equation 13 becomes very simple:

$$q_i/C_i = func_i(q_t) \qquad (14)$$

For the special case of Freundlich equation, when $n_1 = n_2 = \ldots = n$, Equations 7 and 8 give:

$$\pi A/RT = nq_t \tag{15}$$

$$q_i/C_i = K_i^n q_t^{1-n} \tag{16}$$

For the general case of Freundlich equation, when $n_1 \neq n_2 \neq \ldots \neq n_i \neq \ldots$, if the following linear correlation stands:

$$\pi A/RT = \Sigma(n_i q_i) \simeq \alpha q_t \tag{17}$$

Equation 6 becomes:

$$q_i/C_i \simeq (K_i n_i/\alpha)^{n_i} q_t^{1-n_i} \tag{18}$$

These equations — 14, 16, and 18 — indicate that the adsorption equilibrium of a component depends on the total solid-phase concentration.

Simulation for an Imaginary Raw Water

From the result of the simulations given in an earlier section, the correlation of π with q_t and that of q_i/C_i with q_t were plotted in Figure 6 and Figure 7, respectively. These figures verify the simple model given by Equation 14.

Additional simulation was executed for the imaginary water named YX020, YX050, YX150, and YX200 in which the concentration of a certain component (i = 12) is defined to be 1, 2.5, 7.5, and 10 (μmol/L), respectively, and that of other components is kept the same as YA100. The results showed that the correlations given in Figures 6 and 7 do not change. The adsorption isotherm of this component (i = 12) evaluated at a fixed carbon dose is linear as shown in Figure 8 because the value of q_t or π does not change so much. The linear isotherm was also observed when any other component was selected and its initial concentration was slightly changed.

As discussed earlier, the value of q_t on the molar basis is unknown and is measurable only on the basis of a surrogate parameter (q_T) for actual water. Therefore, if there is a correlation between q_t and q_T, Equation 14 can be written as follows:

$$q_i/C_i = func_i(q_T) \tag{19}$$

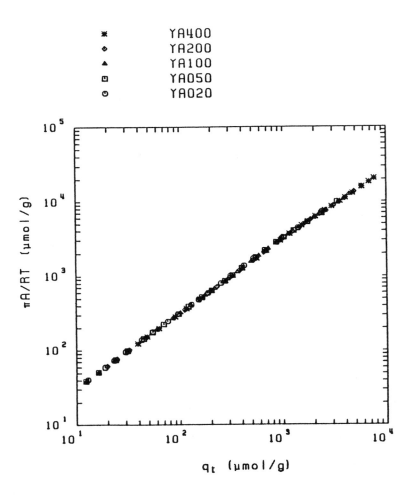

FIGURE 6. Correlation of the spreading pressure and the total solid-phase concentration.

Batch Isotherm of *p*-Aminobenzoic Acid in the Seine River Water

p-Aminobenzoic acid (labeled by ^{14}C) was spiked into the CEB water (the effluent from the slow sand filter which treats the Seine water at the Mont Valerien, near Paris, France). Experimental conditions are shown in Table 2. The sample volume in batch bottles is changed from 100 to 1000 mL according to the carbon (purvelized F400) dose. The results are given in Figures 9 to 11.

The isotherm of the background organics represented by UV-absorbance (at 254 nm, 10-cm cell) is not affected by the slight change of the initial concentration

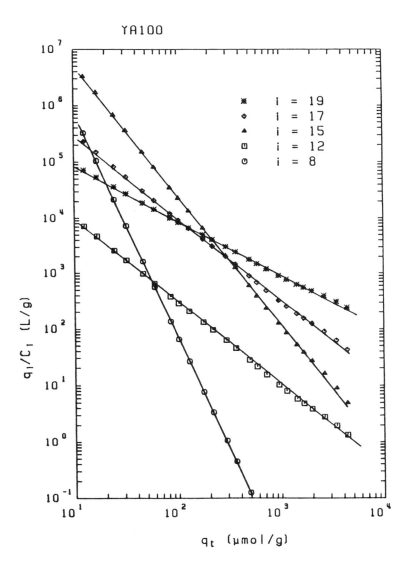

FIGURE 7. Correlation of q_i/C_i and q_i.

of *p*-aminobenzoic acid as shown in Figure 9. The ratio q_i/C_i of *p*-aminobenzoic acid is moderately correlated with q_T (represented by UV-absorbance) as shown in Figure 11 and therefore the isotherm of *p*-aminobenzoic acid evaluated at a fixed carbon dose is almost linear as shown in Figure 10. These results show that the simple model given by Equation 19 is practically acceptable.

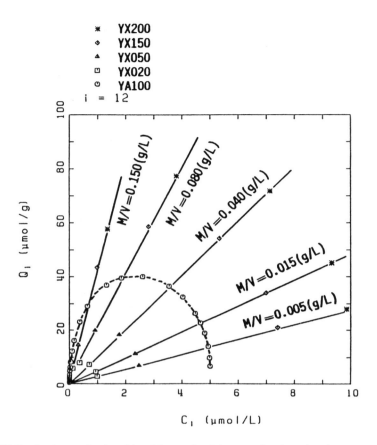

FIGURE 8. Isotherm of solute i (= 12) at a fixed dosage of activated carbon.

Table 2. **Experimental Conditions of Batch Adsorption of *p*-Aminobenzoic Acid in the Seine Water**

	C_{TO} (E_{254})	C_{io} (μmol/L)	M/V (g/L)
Run 1	0.564	0.166	0.005 ~ 0.950
Run 2	0.549 ~ 0.585	0.036 ~ 0.340	0.018
Run 3	0.549 ~ 0.585	0.036 ~ 0.340	0.045
Run 4	0.549 ~ 0.585	0.036 ~ 0.340	0.180

C_{TO} (E_{254}): Initial UV-absorbance at 254 nm, 10-cm cell.
C_{io} (μmol/L): Initial concentration of *p*-nitrobenzoic acid.
M/V (g/L): Dose of activated carbon.

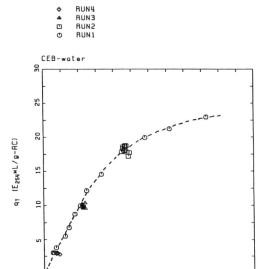

FIGURE 9. Isotherm of the background organics in the Seine water measured by UV-absorbance (at 254 nm, 10-cm cell).

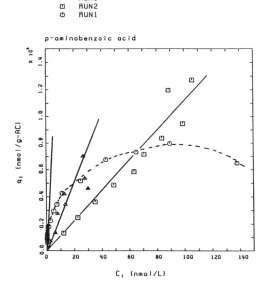

FIGURE 10. Isotherm of *p*-aminobenzoic acid spiked into the Seine water.

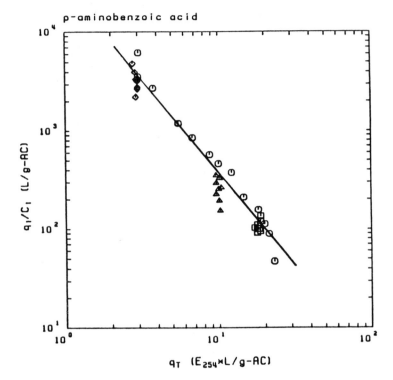

FIGURE 11. Correlation of q_i/C_i of *p*-aminobenzoic acid and q_T of the background organics in the Seine.

CONCLUSION

Two practical models were derived from the results of numerical simulations based on IAS theory. The influence of dilution or concentration of raw water of unknown composition on the isotherm of total organics was described by the modified isotherm model [q_T/q_{TO} or $q_T/C_{TO}^{1/n}$ vs C_T/C_{TO}] which was successfully applied to the experimental batch adsorption equilibria of humic acid measured by UV-absorbance. The second model to describe the influence of background organics upon the adsorption equilibria of a certain component [$q_i/C_i = func_i(q_T)$] was fairly well applied to the experimental adsorption equilibria of *p*-aminobenzoic acid in the Seine River water.

NOTATION

A	specific surface area of activated carbon (cm^2/g)
C_i	liquid-phase concentration of solute i in multisolute system (mol/L)
C_{io}	initial liquid-phase concentration of solute i in multisolute system (mol/L)
C_i^*	liquid-phase concentration of solute i in single-solute system which gives the same value of π as in multisolute system (mol/L)
C_t	total liquid-phase concentration in multisolute system (mol/L)
C_{to}	total initial liquid-phase concentration in multisolute system (mol/L)
C_T	total liquid-phase concentration on the basis of surrogate parameter
C_{TO}	total initial liquid-phase concentration on the basis of surrogate parameter
f_i^*	functions (single-solute isotherms)
$func, func_i$	functions
K, K_i	Freundlich constants
M	mass of activated carbon added (g)
n, n_i	Freundlich constants
N	number of components
q_i	solid-phase concentration of solute i in multisolute system (mol/g)
q_{io}	solid-phase concentration of solute i in multisolute system when V/M is infinitely large (mol/g)
q_{io}^*	solid-phase concentration of solute i in equilibrium with C_{io} in single-solute system (mol/g)
q_i^*	solid-phase concentration of solute i in single-solute system which gives the same value of π as in multisolute system (mol/g)
q_t	total solid-phase concentration in multisolute system (mol/g)
q_{to}	total solid-phase concentration in multisolute system when V/M is infinitely large (mol/g)
q_T	total solid-phase concentration on the basis of surrogate parameter
q_{TO}	total solid-phase concentration on the basis of surrogate parameter when V/M is infinitely large
R	gas constant (erg/mol/K)
T	absolute temperature (K)
V	volume of solution (L)
α	coefficients
ϵ_i	surrogate-parameter/mole conversion coefficient of solute i
π	spreading pressure (dyn/cm)

REFERENCES

1. Myers, A. L. and Prausnitz, J. M. *Am. Inst. Chem. Eng. J.* 11(1):121 (1965).
2. Radke, C. J. and Prausnitz, J. M. *Am. Inst. Chem. Eng. J.* 18(4):761 (1972).
3. Kidney, A. J. and Myers A. L. *Am. Inst. Chem. Eng. J.* 12(5):981 (1966).
4. Jossens, L., J. M. Prausnitz, et al. *Chem. Eng. Sci.* 33:1097 (1978).
5. Singer, P. C. and Yen, C. In: *Activated Carbon Adsorption of Organics from the Aqueous Phase — Vol. 1.* Suffet, I. M. and McGuire, M. J., Eds. (Ann Arbor, MI: Ann Arbor Science, 1980), p. 167.
6. Fritz, W. et al. In: *Activated Carbon Adsorption of Organics from the Aqueous Phase — Vol. 1.* Suffet, I. M. and McGuire, M. J., Eds. (Ann Arbor, MI: Ann Arbor Science, 1980), p. 193.
7. DiGiano, F. A. et al. In: *Activated Carbon Adsorption of Organics from the Aqueous Phase — Vol. 1.* Suffet, I. M. and McGuire, M. J., Eds. (Ann Arbor, MI: Ann Arbor Science, 1980), p. 213.
8. Fritz, W. and Schlunder, E. U. *Chem. Eng. Sci.* 36:721 (1981).
9. Yen C. and Singer, P. C. *J. Environ. Eng.* 110(5):976 (1984).
10. Fettig, J. and Sontheimer, H. *J. Environ. Eng.* 113(4):780 (1987).
11. Crittenden, J. C. et al. *Environ. Sci. Technol.* 19(11):1037 (1985).
12. Crittenden, J. C. et al. *Water Res.* 19(12):1537 (1985).

CHAPTER 3

Using Powdered Activated Carbon: A Critical Review*

I. N. Najm, V. L. Snoeyink, B. W. Lykins, Jr., and J. Q. Adams

Because the performance of powdered activated carbon (PAC) for uses other than taste and odor control is poorly documented, the purpose of this article is to critically review uses that have been reported and to analyze means of employing PAC more efficiently. The extent of adsorption of synthetic organic chemicals on PAC is strongly dependent on the type of compound being removed. The reported removals of trihalomethanes and trihalomethane precursors by PAC range from poor to very good. In selecting the point of addition of PAC, consideration must be given to the degree of mixing, the contact time between the PAC and the water, the PAC residence time, and the minimization of interference of adsorption by treatment chemicals. One of the main advantages of PAC is its low capital cost.

Synthetic and natural organic contaminants are often found in drinking water sources. Categories of these compounds include taste- and odor- (T & O) causing compounds, synthetic organic chemicals (SOCs), pesticides, herbicides, color, and trihalomethane (THM) precursors. The low removals of many undesirable compounds by coagulation, sedimentation, filtration, and chemical oxidation require that an adsorption process be used. Granular activated carbon (GAC) and powdered activated carbon (PAC) can both be used to remove organic compounds from water. Powdered activated carbon is widely used in the United States for T & O removal,[1,2] and GAC filter-adsorbers are commonly used for

---

* Reprinted from *J. Am. Water Works Assoc.*, Vol. 83, No. 1 (January, 1991), by permission. Copyright 1991, American Water Works Association.

the same purpose.³ Granular activated carbon has been the adsorbent of choice for removing SOCs, including volatile organic chemicals (VOCs). There are inherent advantages to applying activated carbon in fixed beds, including lower carbon usage rates for many applications and ease of spent carbon regeneration. However, PAC has the advantages of being a cheaper material and of requiring minimal capital expenditure for feeding and contacting equipment; also, it can be applied only when needed.

In general, the performance of PAC in drinking water treatment for uses other than T & O control is poorly documented. The purpose of this article is to critically review those uses that have been reported and to analyze means of using PAC more efficiently. With this information, it should be possible to determine more easily those situations in which PAC can be used more effectively than GAC.

CHARACTERISTICS OF PAC

The primary characteristic of PAC that differentiates it from GAC is its particle size. Typically, 65 to 95% of commercially available PAC passes through a 325-mesh (44-μm) sieve. For comparison, Kruithof et al.[4] gave the particle size distribution of two powdered activated carbons available in the Netherlands that show 23 to 40% by mass smaller than 10-μm diameter and 10 to 18% larger than 74 μm. The manufacturers' specifications, including particle size distribution, of several PACs commercially available in the United States are summarized in Table 1. The particle size distribution is important because the smaller PAC particles adsorb organic compounds more rapidly than larger particles.[5]

Powdered activated carbon is made from a variety of materials including wood, lignite, and coal. Its apparent density, ranging from 0.36 to 0.74 g/cm³ (23 to 46 lb/cu ft), depends on the type of material and the manufacturing process. Iodine number and molasses number are often used to characterize PAC (Table 1). For example, the AWWA standard for PAC specifies a minimum iodine number of 500.[6]

FACTORS AFFECTING PERFORMANCE OF PAC

Several design and operational parameters affect the performance of PAC for drinking water treatment. Important criteria for selecting the point with addition of PAC include (1) the provision of good mixing or good contact between the PAC and all the water being treated, (2) sufficient time of contact with adsorption of the contaminant, (3) minimal interference by treatment chemicals with adsorption on PAC, and (4) no degradation of finished water quality.

The PAC must be added in a way that ensures its contact with all of the flow. Addition at locations other than those listed in Table 2 may not achieve this objective.

Table 1. Manufacturers' Specifications of Some Commercially Available PACs

Parameter	PAC 1*	PAC 2**	PAC 3***	PAC 4†	PAC 5††	PAC 6†††
Iodine number (mg/g)	800	1,199	600	900	1000	550
Molasses decolorizing index	9			14	18	
Moisture as packed (%)	5	3	5	10	10	4
Apparent density (g/cm³)	0.64	0.54	0.74	0.38	0.38	0.50
Ash content (%)		6		3—5	3—5	
Passing 100 mesh (%)	99		99	95—100	95—100	99
Passing 200 mesh (%)	97		97	85—95	85—95	95
Passing 325 mesh (%)	90	98	90	65—85	65—85	90

* Aqua-Nuchar, Westvaco, Covington, WV.
** WPH, Calgon Corp., Pittsburgh, PA.
*** Aqua, Westvaco, Covington, WV.
† Nuchar S-A, Westvaco, Covington, WV.
†† Nuchar SA-20, Westvaco, Covington, WV.
††† Hydrodarco B, American Norit, Jacksonville, FL.

Table 2. Advantages and Disadvantages of Different Points of Addition of PAC

Point of Addition	Advantages	Disadvantages
Intake	Long contact time, good mixing	Some substances may be adsorbed that would otherwise probably be removed by coagulation, thus increasing carbon-usage rate (this still needs to be demonstrated)
Rapid mix	Good mixing during rapid mix and flocculation, reasonable contact time	Possible reduction in rate of adsorption because of interference by coagulants, contact time may be too short for equilibrium to be reached for some contaminants, some competition may occur from molecules that would otherwise be removed by coagulation
Filter inlet	Efficient use of PAC	Possible loss of PAC to the clearwell and distribution system
Slurry contactor preceding the rapid mix	Excellent mixing for the design contact time, no interference by coagulants, additional contact time possible during flocculation and sedimentation	A new basin and mixer may have to be installed; some competition may occur from molecules that may otherwise be removed by coagulation

Sufficient time of contact is also necessary, and the time required is an important function of the characteristics and concentration of the molecule to be adsorbed.[7] In the absence of competition and coagulation floc interference, 15 min is sufficient time for molecules such as dimethylphenol (mol wt = 122, C_o = 90 mg/L, carbon dose = 250 mg/L) to equilibrate with a 325-mesh (44-μm-diameter) particle if adequate mixing is used.[8] As the molecular size increases, the rate of diffusion into the pores of the PAC particle decreases. For example, rhodamine B dye (mol wt = 422) requires approximately 5 hr to come to equilibrium, a 10,000-mol wt fulvic acid requires 17 hr, and a 50,000-mol wt humic acid requires about 2 days. The adsorption kinetics and equilibrium capacity depend on the type of carbon used and, therefore, these values should only be taken as rough estimates. If insufficient time is allowed for equilibration, more PAC must be used to achieve the desired removal.

The effect of PAC particle size on required contact time in the absence of floc interference was shown by Najm et al.[5] for a continuous stirred tank reactor. In 15 min, 500 μg/L trichlorophenol was reduced to 25 μg/L by 14-μm-diameter PAC but only to 275 μg/L by 100-μm-diameter PAC. The composite PAC sample, with an average diameter of 40 μm, reduced the concentration to 180 μg/L. In no case was equilibrium achieved in 15 min, and the concentration in the test with the composite sample was still decreasing after 120 min.

MEANS OF APPLYING PAC

Powdered activated carbon can be fed as a powder using dry feed machines or as a slurry using metering pumps.[9] In a conventional treatment plant, the common points of PAC addition are the plant intake, rapid mix, and filter influent. Another point of addition that should be considered, although it is not commonly used, is a continuous-flow slurry contactor that precedes the rapid mix. The PAC can be mixed intensely with the water in the absence of floc to obtain rapid adsorption kinetics and then incorporated into the floc in the rapid mix for subsequent removal by sedimentation and filtration.[5] Table 2 summarizes some of the important advantages and disadvantages of PAC addition at each of these points.

When PAC is added at the rapid mix, incorporation of PAC into the coagulant floc particles may reduce the rate of adsorption.[10-12] Gauntlett and Packham[10] conducted jar tests showing that the removal rate of chlorophenol by PAC in the absence of alum addition was most rapid and that the addition of PAC after alum addition gave a better rate of removal than when applied just before alum. Apparently, PAC added after alum floc has formed adhered to the outer surface of the floc rather than being incorporated into the floc. Gauntlett and Packham argued that the reason for the reduction in removal rate is that the adsorbate must diffuse through the part of the floc surrounding the PAC particle and then into the particle itself in order to be adsorbed. Najm et al.[12] however, found little reduction in the rate of adsorption of trichlorophenol on PAC in spite of the incorporation of the carbon particles into coagulant flocs.

Addition of PAC at the intake has the advantage of providing extra contact time, but there is the possibility of the PAC adsorbing many compounds that would otherwise be removed by coagulation, flocculation, and sedimentation. Onsite tests are recommended to determine whether one factor outweighs the other.

Addition of PAC just before the filter is advantageous because the PAC can be retained in the filter and kept in contact with the water longer, thereby better using its capacity. The average PAC residence time is equal to one half of the time between two successive backwashings, assuming that PAC is continuously added to the influent water. However, the PAC must be added carefully to avoid its penetration into the distribution system. The maximum dosage of PAC is limited by the ability of the filter to retain the PAC and by the rate of head loss buildup in the filter, which is expected to increase as PAC dosage increases. Sontheimer et al.[13] reported a 10 to 20% reduction in filter run time as a result of the addition of PAC. Dougharty and Morris[14] reported that filter velocities $\geqslant 8$ to 9 m/hr resulted in a breakthrough of PAC when it was added at a dose of 60 mg/L. In a direct filtration pilot plant study, Gifford et al.[15] reported a 40% reduction in filter run time as a result of the addition of 30 mg PAC per liter. With a hydraulic loading of 4.9 m/hr, no breakthrough of PAC was observed in the filter effluent.

Sontheimer[11] suggested adding a separate PAC reactor between the sedimentation basin and the filter to increase the time of contact and take advantage of coagulation and sedimentation for eliminating competing organics to the maximum extent possible. A major disadvantage, however, is that an additional coagulation step would be required to remove the PAC.

Improving Adsorption Efficiency

Various techniques for applying PAC to improve its adsorption of large, slowly diffusing compounds are available. For example, adding PAC to solids contact clarifiers has the potential for improved adsorption efficiency because the carbon can be kept in contact with the water for a longer time than when it is added to the rapid mix of a conventional plant. Kassam et al.[16] reported mean carbon residence times ranging from 9 hr to 8.5 days when PAC was added to a solids contact slurry recirculating clarifier. Hoehn et al.[17] studied the addition of PAC to a pilot-scale floc-blanket clarifier. They reported PAC retention times of 1 to 2 days. In a study of the addition of PAC to a bench-scale floc-blanket reactor, Najm et al.[12] measured steady-state PAC retention times ranging from 9 hr to 2 days. They determined that the maximum adsorptive capacity of PAC for 2,4,6-trichlorophenol (TCP) was achieved during this time. Richard[18] noted that the PAC dosage for detergent removal could be reduced by 25 to 40% if the carbon was added to the influent of a floc-blanket clarifier instead of to a conventional system. More research is needed to optimize this process and to determine coagulation procedures that can be used to increase adsorption kinetics, such as using a polyelectrolyte alone instead of alum or ferric salts.[8,19]

In the Roberts-Haberer process,[20-22] buoyant polystyrene spheres 1 to 3 mm in diameter are coated with PAC. The spheres are held in the reactor by means of a screen, and the water to be treated is passed upflow through the medium. After saturation, the PAC is removed from the beads by backwashing the medium (downflow) with a high flow rate. New PAC is then applied to the beads. The PAC at the entrance to the process can be equilibrated with the influent concentration, thereby increasing its adsorption efficiency. Recovery of the PAC for regeneration may also be possible. In addition, the Roberts-Haberer filter may be used as a roughing filter to remove some suspended solids before filtration. However, careful studies are needed to assess the efficiency of adsorption for specific organics and the ability of the polystyrene beads to retain the PAC in order to determine the conditions under which the process will be more cost-effective than GAC.

Sontheimer et al.[23] suggested the addition of PAC to a multistage contactor. They showed that addition of two doses of PAC, m_1 and m_2, to contactors 1 and 2, respectively, in series theoretically results in a lower usage rate than addition of the total carbon dose, $m_T = m_1 + m_2$, to one contactor. To determine the minimum values of m_1 and m_2 required to drop the adsorbate concentration from an initial concentration, C_o, to a final concentration, C_2, the intermediate aqueous concentration, C_1, between the two reactors was calculated. Assuming that the PAC reached equilibrium with the effluent of each contactor and that the same Freundlich isotherm equation describes equilibrium in both contactors, C_1 was calculated by

$$\left(\frac{C_1}{C_2}\right)^{1/n} - n\frac{C_0}{C_1} = 1 - \frac{1}{n} \qquad (1)$$

in which C_o = influent concentration to contactor 1, C_1 = effluent concentration of contactor 1 and influent concentration to contactor 2, C_2 = effluent concentration of contact 2, and $1/n$ = Freundlich equation constant.

The PAC dose to each reactor can then be calculated using Equation 2, described later. The limitation to this approach is that equilibrium is not always reached when PAC is added to conventional contactors and the same Freundlich equation may not apply in each contactor. When competing organics are present, the isotherm function depends on PAC dosage.

Reactions with Treatment Chemicals and Calcium Carbonate

Careful attention must be paid to the interaction of PAC with water treatment chemicals. Activated carbon will chemically reduce compounds such as free and combined chlorine, chlorine dioxide, chlorite, ozone, and permanganate; the demand for oxidants and disinfectants is thereby increased. The reaction of activated carbon with chlorine will reduce its adsorptive capacity for compounds such as phenol and substituted phenols.[24,25] Lalezary-Craig et al.[26] found a reduction in the ability of PAC to adsorb both geosmin and methylisoborneol in water containing free chlorine and monochloramine. Surprisingly, the effect of the monochloramine appeared to be greater than that of chlorine.

Addition of PAC to a water that is supersaturated with calcium carbonate or other precipitates, or treatment that causes an increase in pH to greater than the saturation pH just after PAC is added, such as in lime softening or aeration of water with a high carbon dioxide content, may lead to coating of the PAC particle with precipitates and to a corresponding decrease in adsorption efficiency. Also, adsorption at high pH is often poorer than at low pH because many organic contaminants are weak acids that ionize at high pH, rendering them more hydrophilic.

PAC PERFORMANCE

Many studies have been conducted to assess the adsorption of organic compounds on PAC under both laboratory and field conditions. Various types of organics, activated carbons, and waters were included in these studies. The most common means of evaluating adsorption on activated carbon, especially PAC, has been through the use of adsorption isotherms. The experimental bottle-point technique is widely used to achieve an isotherm relationship between the aqueous adsorbate concentration and the carbon surface loading (see Randtke and Snoeyink[9] for a description of this test). Because these isotherms are conducted under laboratory conditions, the accuracy of performance predictions based on them needs to be examined.

Table 3 lists some of the compounds for which isotherms have been determined along with their Freundlich isotherm constants. Also listed are the predicted carbon dosages required to reduce the concentration of each of these compounds from 100 to 10 and 1 µg/L, and from 10 to 1 µg/L. Each of these calculations used the isotherm of that compound and assumed that equilibrium is achieved at the final concentration. The isotherm will change as the type of carbon and the type of water treated change. The following equation was used to calculate the carbon dosages required:

$$m = \frac{(C_0 - C_\ell)}{q_\ell} = \frac{(C_0 - C_\ell)}{KC_\ell^{1/n}} \qquad (2)$$

in which m is the carbon dosage (mg/L), C_0 and C_e are the initial and equilibrium concentrations (µg/L), respectively, and K and n are the Freundlich equation constants.

All the isotherm data listed in Table 3, except those of Lalezary et al.[27] and Najm et al.,[12] were conducted using GAC pulverized to a smaller size to enhance kinetics and reduce the time required for equilibration. Some of the isotherms of Miltner et al.[28] were conducted with 50 × 200-mesh pulverized GAC (74-297-µm diameter), whereas the rest of the isotherms were conducted using pulverized GAC of <200 mesh.

A major concern with some of the isotherm data listed in Table 3 relates to the contact time allowed between the carbon and the water before sampling. Although higher adsorption rates are achieved when GAC is pulverized, some large molecular weight organic compounds may not reach equilibrium in less than 1 or 2 days. The isotherms for large molecules determined with 1 hr or less of contact time can be used as rough estimates of adsorbability, but probably indicate a capacity lower than achievable at equilibrium.

Caution should be exercised when using the bottle-point isotherm of a trace compound conducted with natural water. The competitive effect of the background natural organic matter on the adsorption isotherm of the trace compound on activated carbon has to be carefully considered when the isotherm is being used to predict the performance of an activated carbon process. When conducted in a multisolute system, the bottle-point isotherm of a trace compound depends on the initial concentration used. Frick[29] determined that the isotherm capacity of activated carbon for *p*-nitrophenol (PNP) decreased with decreasing initial

Table 3. Isotherms for Selected Organic Contaminants

Compound	Water*	K**	1/n	PAC Dose (mg/L) $C_o = 100$ µg/L $C_e = 10$ µg/L	PAC Dose (mg/L) $C_o = 100$ µg/L $C_e = 1$ µg/L	PAC Dose (mg/L) $C_o = 10$ µg/L $C_e = 1$ µg/L	PAC Dose (mg/L) $C_o = 10$ µg/L $C_e = 10$ µg/L (sic)	Time***	Range†	Ref.
Alachlor	DDW	80.2	0.26	0.6	1.2		0.1	14 days	512—0.6	28
	GW	62	0.33	0.7	1.6		0.14	18 days	477—0.2	28
	SW	10.5	0.38	3.6	9.4		0.86	11 days	46—2.6	28
Aldicarb	DDW	8.4	0.4	4.3	11.8		1.1	13 days	274—1.3	28
	SW	4.52	0.41	7.7	21.9		2.0	12 days	171—19	28
Atrazine	DDW	38.2	0.29	1.2	2.6		0.24		785—4.2	28
	GW	24.4	0.36	1.6	4.0		0.37		317—0.6	28
Benzene	MDW	1.4E-9	2.9	>100,000	>100,000		>100,000	1 hr	8,000—3,000	47
		3.57	0.42	9.6	27.7		2.5	15 min	50,000—3,000	72
Bromodichloromethane	MDW	0.10	0.76	156.4	990		90	1 hr	57—1.3	47
Bromoform	MDW	0.65	0.83	20.5	152.3		13.8	1 hr	90—2	47
Carbofuran	DDW	16.2	0.41	2.2	6.1		0.6	7 days	111—2.9	38
	SW	11.9	0.36	3.3	8.3		0.76	7 days	365—2.2	38
Carbon tetrachloride	MDW	0.12	0.84	108.4	825		75	1 hr	120—2.5	47
Chlorobenzene	MDW	0.11	0.98	85.7	900		81.8	1 hr	17,500—1,400	47
	DDW	9.0	0.35	4.5	11.0		1.0	11 days	732—15.4	28
	SW	9.9	0.31	4.4	10.0		0.9	10 days	401—14.1	28
Chloroform	MDW	0.03	0.84	433.6	3,300		300	1 hr	75—5	47
Cyclohexane	MDW	23.9	0.2	2.4	4.14		0.4	15 min	50,000—3,000	72
Dibromochloromethane	MDW	0.10	0.93	105.7	990		90	1 hr	40—1.5	47
Dibromochloropropane	DDW	6.61	0.51	4.2	15.0		1.4	8 days	69.2—0.2	28
	SW	3.84	0.46	8.1	25.8		2.3	8 days	85.8—0.2	28
o-Dichlorobenzene	DDW	19.1	0.38	2.0	5.2		0.5	11 days	1,056—14.4	28
	SW	8.4	0.58	2.8	11.8		1.1	10 days	293—59	28
cis-1,2-Dichloroethylene	DDW	0.2	0.59	115.7	495		45	3 days	615—5.5	28
	SW	0.15	0.64	137.4	660		60	31 days	230—19	28
	GW	0.12	0.63	176	825		75	15 days	214—9.7	28
trans-1,2-Dichloroethylene	DDW	0.62	0.45	51.5	160		14.5	10 days	415—13.5	28
	SW	0.63	0.39	58.2	157		14.3	10 days	406—59	28

Table 3 (continued). Isotherms for Selected Organic Contaminants

Compound	Water*	K**	1/n	PAC Dose (mg/L)				Time***	Range†	Ref.
				$C_o = 100$ μg/L		$C_o = 10$ μg/L				
				$C_e = 10$ μg/L	$C_e = 1$ μg/L	$C_e = 1$ μg/L	$C_e = 1$ μg/L			
2,4-Dichlorophenol	GW	23.3	0.29	2.0	4.2	0.4		7 days	300—1	12
	SW	13.4	0.32	3.2	7.4	0.67		30 min	600—50	35
1,2-Dichloropropane	DDW	0.3	0.59	77.1	330	30		3 days	196—4.1	28
	SW	0.4	0.48	74.5	247.5	22.5		10 days	126—4.6	28
Dieldrin	DDW	3.0	0.56	8.3	33	3				41
Endrin	DW	5.0	0.85	2.5	19.8	1.8				41
2,4,5-T ester	DDW	3.6	0.79	4.0	28.5	2.5				41
Ethylbenzene	DDW	0.23	0.79	63.5	430.4	39.1		1 hr	22,500—2,100	47
	MDW	16.2	0.29	2.8	6.11	0.56		15 min	50,000—3,000	72
	DDW	4.5	0.53	5.9	22	2		3 days	565—5.6	28
Ethylene chloride	MDW	0.0011	1.5	2,587	90,000	8,182		1 hr	65—17	47
Ethylenedibromide	DDW	0.9	0.46	34.7	110	10		4 days	123—0.2	28
	SW	0.77	0.54	33.7	128.6	11.7		10 days	139—1.1	28
Geosmin	DDW	13.5	0.39	2.7	7.33	0.67		5 days	0.05—2.0	27
	DDW	0.18	0.83	74.0	550	50		5 days	0.008—0.05	27
Heptachlor	DDW	1.78	0.92	6.1	55.6	5.1			40—0.16	71
Heptachlor epoxide	DDW	11.92	0.75	1.34	9.6	0.76			30—0.04	71
Hexachlorobutadiene	MDW	4.64	0.63	4.5	21.3	1.90		1 hr	117—1.3	47
Lindane	DDW	4.5	0.74	3.6	22	2.0				41
	DDW	15.4	0.43	2.2	6.4	0.6		6 days	170—2.1	28
	SW	0.31	0.34	132.7	319.4	29				41
	SW	14.5	0.39	2.5	6.8	0.6		7 days	149—0.6	28
Methoxychlor	DDW	10.0	0.36	3.8	9.9	0.9			460—0.016	28
	TW	28.0††	0.17	2.16	3.53	0.32			1,000—100	74
	TW	5.83†††	0.46	5.4	17.0	1.5			500—40	74
Methylisoborneol	DDW	5.03	0.5					5 days	0.04—5	27
	DDW	169.3	1.65					5 days	0.02—0.04	27
Parathion	DDW	3.6	0.26	13.7	27.5	2.5				41
PCB (Aroclor 1254)	DDW	11.46	1.03	0.73	8.6	0.8		28 days	0.17—0.011	28
	HA	2.7	0.99	3.4	36.7	3.33		28 days	0.8—0.058	28

Pentachlorophenol	MDW	8.24	0.42	4.2	1.1	1 hr	8,180—220	47
	DDW	42.4	0.34	1.0	0.2	7 days	3,408—0.4	28
	SW	14.4	0.36	2.7	0.6	7 days	322—5.3	28
Phenol	MDW	0.62	0.51	44.8	14.5	1 hr	9,360—2,150	47
Simazine	DDW	30.8	0.23	1.7	0.3		16.5—0.02	38
	GW	22.3	0.31	2.0	0.4		177—0.2	38
Styrene	DDW	12.1	0.48	2.5	0.7	12 days	148—11.8	28
Toluene	DDW	15.9	0.28	3.0	0.57	15 min	50,000—3,000	72
	DDW	4.5	0.45	7.1	2	4 days	104—2.3	28
	GW	6.3	0.37	6.1	1.4	15 days	672—2.5	28
	GW	8.72	0.45	3.7	1.0	30 min	180—14	73
Toxaphene	DDW	5.725	0.74	2.9	1.57	28 days	0.5—0.064	28
	HA	2.97	0.81	4.7	3.0	28 days	2.2—0.32	28
Trichloroethylene	MDW	0.66	0.50	43.1	13.6	1 hr	63—0.15	47
	DDW	1.35	0.51	20.6	6.67	7 days	31—2.88	12
	GW	0.9	0.51	30.9	10.0	7 days	32—3.9	12
	DDW	1.0	0.9	11.3	9.0		1,000—6	23
	TW	1.0	0.9	11.3	9.0		1,000—6	23
2,4,6-Trichlorophenol	GW	10.3	0.32	4.2	0.87	7 days	350—10	12
	DDW	21.2	0.3	2.13	0.4	7 days	400—4	12
	DDW	1.3	0.75	12.3	6.9	3 days	166—5.6	28
m-Xylene		20.13	0.27	2.4	0.45	15 min	50,000—3,000	72
o-Xylene	DDW	7.1	0.47	4.3	1.3	8 days	1,795—6.9	28
	GW	9.9	0.47	3.1	0.9	10 days	522—10.5	28
p-Xylene	MDW	28.15	0.16	2.2	0.3	1 hr	12,000—1,500	47
	DDW	11.1	0.42	3.1	0.8	9 days	32.7—1.6	28
	GW	7.6	0.46	4.1	1.2	15 days	232—2.9	28

* MDW — mineralized distilled water, DDW — distilled deionized water, GW — ground water, SW — surface water, TW — tap water, HA — humic acid (5 mg/L as TOC).
** Freundlich K values are listed for C in μg/L and in mg/g.
*** Contact time in the isotherm bottle.
† Range of remaining aqueous concentration in μg/L.
†† C_o = 5,000 μg/L.
††† C_o = 1,000 μg/L.

PNP concentration in the presence of a constant background concentration of p-hydroxybenzoic acid (HBA) and phenol-4-sulfonic acid (PSA). With an initial concentration of 5 mg/L of each of HBA and PSA, a decrease in initial PNP concentration from 10 to 2 mg/L resulted in a 50% reduction in the carbon's capacity for PNP. Similarly, the authors' research showed that the carbon capacity for TCP present in natural water and determined using the bottle-point isotherm technique depended on the initial concentration of TCP. At a certain TCP equilibrium concentration, the carbon capacity decreased with decreasing initial concentration as a result of the competitive effect of the background natural organics.

Therefore, the choice of the proper isotherm significantly affects the evaluation of the performance of PAC in a continuous process. Although the effect of competitive adsorption behavior on the kinetic models is not clearly understood, the authors believe that the isotherm to be used should be that conducted with an initial concentration close to the concentration of the influent to the continuous process.

Published reports of PAC performance are discussed subsequently and, when possible, are compared with performance predictions based on equilibrium isotherms. The data on PAC performance are categorized according to type of compound as well as according to impact on drinking water quality.

T & O-Causing Compounds

Powdered activated carbon is widely used in the United States for T & O control. It has been estimated that about 90% of water utilities worldwide that use activated carbon use it in the form of PAC.[11,30] In 1977 PAC was used in 25% of the 683 U.S. water utilities surveyed, which included the 500 largest utilities.[1] The number of utilities using PAC is constantly on the rise.[2] Compounds causing T & O problems vary widely from the strongly adsorbed, low-molecular-weight compounds, such as 2,4-dichlorophenol (2,4-DCP), to the weakly adsorbed, high-molecular-weight compounds, such as some types of humic substances. This has resulted in a wide variation of PAC adsorption efficiency.

The concentration of organic compounds that cause a T & O problem is highly dependent on the type of compound. The threshold odor number (TON) is widely used as a measure of taste and odor in drinking water. The TON produced by a given concentration of contaminant is usually different for each compound. For example, geosmin is known to cause odor problems in the low nanograms-per-liter range of concentrations,[27] whereas 2 μg/L of PNP causes objectionable odor and 8 μg/L causes a taste problem.[31] Sigworth[32] presented a list of T & O-causing herbicides and pesticides along with the concentrations that caused detectable odor. The concentrations varied from 17 μg/L for aldrin to 4.7 mg/L for methoxychlor.

The concentration of PAC required to remove T & O varies widely, consistent with the diversity of T & O-causing compounds. Researchers[33] found that 1 mg

PAC per liter reduced the TON by 1 to 24 units in the water supplies of four U.S. cities, indicating a wide variability in adsorbability of T & O-causing compounds. Bench-scale studies of three types of carbon showed that 75 mg/L of the best carbon was required to achieve humic odor reduction from a TON of 15 to 3 and 30 mg/L of the best PAC was required to reduce the concentration of 2,4-DCP from 1.75 to 0.24 mg/L, corresponding to a TON reduction from 22 to 3.[34] The isotherms of Najm et al.[12] and Aly and Faust[35] (see Table 3), however, predict PAC doses of 13 and 20 mg/L, respectively, to achieve the same removal of 2,4-DCP.

A full-scale study was conducted on the performance of PAC for the removal of T & O at the East Drive Well in North Miami Beach, FL.[36] Doses of PAC up to 27 mg/L failed to remove a hydrogen-sulfide-like odor. In another full-scale study,[37] an average PAC dose of 23 mg/L reduced the TON of influent water to the Nitro, WV, water plant from 30 to 325 to 14.6, which was still much higher than the acceptable value of 3 or less. The specific compounds causing the T & O were not determined, so comparison with predicted dosages based on the isotherm is not possible.

Laboratory tests were conducted by Lalezary et al.[27] to study the efficiency of two types of PAC (PAC 1* and PAC 2**) for the removal of the following odor-causing compounds — geosmin, 2,3,6-trichloroanisole (TCA), 2-isopropyl-3-methoxypyrazine, 2-isobutyl-3-methoxypyrazine, and 2-methylisoborneol (MIB) — at the nanograms per liter concentration level. The effects of chlorination, coagulation, and filtration on the adsorption of these compounds on the two types of PAC were investigated. In a pilot study that followed, Lalezary-Craig et al.[26] investigated the effects of initial concentration, chlorine and chloramine residual, presence of background organics, contact time, and filtration rate on the removal of geosmin and MIBB by PAC 2. The authors reported that carbon doses as low as 10 mg/L reduced 66 ng/L each of geosmin and MIB to 2 and 7 ng/L, respectively, for surface coverages of approximately 6 µg/g. However, data are not available in this range to determine whether equilibrium had been achieved. The presence of chlorine and chloramine, as well as background organics, was found to decrease the removal of the two compounds by PAC. The removals of geosmin and MIB achieved by PAC in the pilot plant compared well with those predicted by the bench-scale studies for similar initial concentration, chlorine dosage, and coagulation conditions.[26,27]

Pesticides and Herbicides

A growing problem facing many water treatment facilities is the increasing public concern, and the related regulations, regarding the concentrations of

---

* Aqua Nuchar, Westvaco Corp., Covington, WV.

** WPH, Calgon Corp., Pittsburgh, PA.

pesticides and herbicides (collectively termed biocides) in drinking water sources. Biocides are transported from agricultural fields into lakes and rivers by water runoff. Miltner et al.[38] reported the presence of several biocides in the Sandusky River in northwestern Ohio during a survey from May to July 1984. These biocides included alachlor, metolachlor, atrazine, cyanazine, metribuzin, carbofuran, and simazine at mean concentrations ranging from 0.44 μg/L for simazine to 8.17 μg/L for metolachlor. The maximum concentrations, however, ranged from 1.15 to 22.3 μg/L.

Six biocides — endrin, lindane, methoxychlor, toxaphene, 2,4-D, and 2,4,5-TP — are included in the current drinking water standards.[39] New standards, proposed in May 1989,[10] do not include endrin, but include 11 additional biocides. Table 3 lists isotherm data for some of these compounds, showing them to be strongly adsorbed. Table 3 shows that a PAC dose of <5 mg/L should be capable of reducing the concentration of any of the biocides listed from 10 to <1 μg/L and that the type of water affects the isotherm capacity, an effect probably caused by natural background organics.

A comparison of the carbon dosages required to reduce the pesticide concentration to a predetermined value in a jar test using distilled and natural water is shown in Table 4.[41] The data show that the adsorptive capacity of PAC in raw river water is 50 to 93% lower than that in distilled water.

Miltner et al.[38] investigated the removal of several pesticides that were included in a national pesticide survey. Alachlor, atrazine, carbofuran, linuron, metolachlor, metribuzin, and simazine were found in the influent water to three water treatment plants in Ohio. The performance of PAC for their removal was investigated in full-scale as well as bench-scale studies. The isotherms for these compounds on pulverized GAC are listed in Table 3. Although the adsorption capacities of PAC in jar tests and full-scale treatment were in good agreement, they were approximately two orders of magnitude lower than those predicted by the pulverized GAC isotherms listed in Table 3. One possible reason for this difference is that the initial concentrations in the isotherm studies were much higher than those used in the jar tests and full-scale studies. This would lead to a greater reduction in biocide adsorption capacity in the jar tests and full-scale plants, as discussed earlier in this article. Other possible reasons for the difference include (1) the short contact time allowed (1.5 hr* compared with several days for the pulverized GAC isotherms) and (2) some interference by coagulants.

The study of Miltner et al.[38] concluded that all pesticides detected in the Sandusky and Maumee rivers in Ohio will not be removed by conventional water treatment processes unless an activated carbon adsorption process is employed. Similar conclusions were reached by Richard et al.[42] for the removal of atrazine from the Des Moines, IA, water supply. Poor removals of the pesticide atrazine by PAC, however, were observed at New Orleans, LA.[43,44]

---

* The total contact time included a period of sedimentation when mixing was not employed. The amount of adsorption that occurred during this time was not established.

Table 4. PAC Dose Required to Reduce the Pesticide Level in Distilled and Little Miami River Water[11]

Pesticide	Method	PAC Dose (mg/L)			
		$C_o = 10$ µg/L		$C_o = 1.0$ µg/L	
		$C_e = 1.0$ µg/L	$C_e = 0.1$ µg/L	$C_e = 0.1$ µg/L	$C_e = 0.05$ µg/L
Parathion	JT*	2.5	5	0.5	0.6
	P**	5	10	0.9	1.1
2,4,5-T ester	JT	2.5	17	1.5	3
	P	14	44	3	5
Endrin	JT	1.8	14	1.3	2.5
	P	11	126	11	23
Lindane	JT	2	12	1.1	2
	P	29	70	6	9
Dieldrin	JT	3	12	1.1	1.7
	P	18	85	7	12

* PAC dosage in jar test in which pesticide is removed from distilled water by PAC alone, with a contact time of 1 hr.
** PAC dosage in the plant in which pesticide is removed from river water by conventional treatment and activated carbon.

SOCs

Synthetic organic compounds are the products of a wide range of manufacturing processes and some are classified as hazardous to animal and human health. Occasionally, these compounds find their way into natural water sources by means of accidental spills, dumping, or leaching to groundwater aquifers. The U.S. Environmental Protection Agency (U.S. EPA) has set maximum contaminant levels (MCLs) for several VOCs,[45] and many other compounds are currently being added to the list (see the *Federal Register*[46] for the proposed listing). Many of these compounds are included in Table 3.

The adsorbability of SOCs is highly dependent on the type and properties of the compound being removed. A bench-scale study on PAC adsorption of four SOCs present in Mississippi River water was conducted at New Orleans, LA.[43] The removals of 1,2-dichloroethane, benzene, toluene, and atrazine by conventional treatment processes (coagulation, flocculation, sedimentation, filtration, and chlorination) were investigated, along with their adsorption on 5, 50, and 500 mg PAC per liter. The raw water concentrations of the compounds, however, never exceeded 1 μg/L and continuously varied throughout the period of the study. These facts made it difficult to draw solid conclusions concerning the removal of these compounds. The data, however, suggest that removal did not occur unless PAC was present.

The removal of TCP and trichloroethylene (TCE) from groundwater by PAC added to a bench-scale floc-blanket reactor was investigated by Najm et al.[12] With the long carbon residence times supplied (1 to 2 days), complete utilization of the adsorptive capacity of the carbon, as indicated by the isotherm, was achieved.

A major spill of carbon tetrachloride (CCl_4) in a tributary of the Ohio River during early 1977 resulted in several studies of its removal by PAC. Bench-scale studies at the U.S. EPA laboratory in Cincinnati utilized untreated river water containing approximately 70 μg CCl_4/L. The water was dosed with varying amounts of PAC, mixed rapidly for 1 min, and then contacted for 1 hr in a jar. The highest PAC dose of 30 mg/L resulted in CCl_4 reduction of only 20%. The amount adsorbed on the carbon surface, 1.86 mg/g, was higher than the equilibrium capacity predicted by the 1-hr isotherm of Dobbs et al.[47] given in Table 3. During this period, 9.6 mg PAC per liter was used at the Cincinnati, OH, water treatment plant. Using conventional treatment and this PAC dosage, the CCl_4 concentration in the finished water was virtually equal to that in the influent water to the plant.[48]

Following the CCl_4 incident, another spill of toxic chemicals occurred in the same river at a different location. The spill consisted of hexachlorocyclopentadiene and octachlorocyclopentadiene. Although the concentrations of the compounds were quite low (<1 μg/L), up to 30 mg PAC per liter was added as a precaution at the Evansville, IN plant.[49] Both compounds were virtually absent (<0.1 μg/L) in the finished water.

After consumers' complaints of a pesticidelike taste and odor in the drinking water of the Sunny Isles water treatment plant, North Miami Beach, FL, in September 1977, laboratory analysis of the raw and finished water revealed the presence of 42 SOCs ranging in concentration from 0.01 to 73 µg/L.[36,50,51] A study of the removal of natural total organic carbon (TOC) and SOCs by 7.5, 15, and 30 mg PAC per liter was initiated. The PAC was added at the well intake to achieve a maximum of 2 hr of contact time. A finished water concentration of 40 µg/L of volatile halogenated aliphatics, including dichloroethenes and chloroethanes (1,1-dichlorethane and 1,1,1-trichloroethane), was virtually unaffected by the addition of up to 26 mg PAC per liter. The chlorobenzene concentration in the finished water was reduced from 0.5 to <0.1 µg/L with the addition of 10 mg PAC per liter. Nonvolatile synthetic organic chemicals were reduced from a combined finished water concentration of 45 to <5 µg/L with the addition of 7.5 mg PAC per liter. Comparison of the removal data, however, with those predicted by the isotherms listed in Table 3 was not possible because the concentrations of the compounds were far lower than those used in the isotherm tests.

THMs and THMFP

The discovery that THMs are formed during drinking water treatment[52,53] has caused major concern because of the potential health hazard associated with these compounds. Rook[52] attributed the formation of the compounds to the reaction between natural organic matter and chlorine. Organic compounds that undergo this reaction are referred to as THM precursors. The amount of THMs that will be produced if the reaction goes to completion is referred to as the THM formation potential (THMFP). In 1979 the U.S. EPA established an MCL for total THMs at 100 µg/L as an annual average.[54] This MCL is currently being reviewed by the U.S. EPA.

Adsorption of THM Precursors

The reduction of THMFP can be determined by measuring THMFP directly or by measuring TOC. Although there is usually a good correlation between the TOC and THMFP for a given type of water, there is no correlation that applies to all waters because of the differences in the characteristics of the organics. For example, Lange and Kawczynski[55] observed significant removal of TOC with virtually no reduction in THMFP. Hentz et al.[56] observed similar results for another water. Amy and Chadik,[57] however, reported a substantially greater reduction of THMFP than of TOC with the addition of PAC.

During conventional water treatment, part of the natural organic matter comprising the THMFP is removed by chemical coagulation. Although these removals are usually low,[55] some have been reported to be as high as 86%.[58] Amy and Chadik[57] reported a reduction in the THMFP of Ilwaco Reservoir (WA) water from 1,087 to 371 µg/L with an alum dose of 2.5 mg/L (a 66% reduction) and to 245 µg/L with an optimum ferric dose of 7.5 mg/L (a 77% reduction).

The removal of THM precursors by adsorption on activated carbon has varied from poor to very good. Many studies have reported the need for very high PAC dosages to achieve reductions of THMFP. In a pilot-scale study of conventional treatment processes, Love et al.[59] investigated the reduction of THMFP by PAC (Table 5). A PAC dose of 100 mg/L was required to reduce the THMFP from 53 to 25 µg/L. In a bench-scale study at New Orleans, LA,[43] researchers determined that 90% reduction of THMFP could be achieved, compared with that obtained by conventional treatment, with a PAC dose as high as 500 mg/L (Table 6). Blank runs of conventional treatment with no PAC added resulted in THMFP reduction efficiencies varying from 27% in the spring to 54% in the fall. Hentz et al.[56] determined that PAC doses ranging from 75 to 125 mg/L were required to achieve a slight reduction of 7.5 to 29% of a THMFP of approximately 425 µg/L. Their results agree with those of Wood and Demarco,[60] Symons,[61] and Lange and Kawczynski,[55] who also found PAC to be ineffective for the removal of THM precursors.

The use of PAC was evaluated for the removal of chloroform precursors in a study involving 15 of Kentucky's larger water utilities.[62] After the Ohio River water had been chlorinated in a jar with a dose of 2 to 3 mg chlorine per liter and then stored for 3 days, the terminal chloroform concentration was 236 µg/L (Table 7). As shown in Table 7, a PAC dose of 50 mg/L applied prior to chlorination resulted in a 50% decrease in the terminal chloroform concentration. With a PAC dose of 1000 mg/L, however, 20% of the chloroform formation potential (47.5 µg/L) was still not reduced. Addition of PAC to a pilot-scale floc-blanket reactor for the control of THMFP was investigated by Hoehn et al.[17] The addition of 25 mg/L of PAC* to water containing 362 µg THMFP per liter improved the reduction efficiency by a mere 4% over the 56% reduction achieved by coagulation alone. However, the addition of 21 mg/L of another PAC,** a higher-grade carbon, resulted in an added 16% reduction in THMFP.

The PAC adsorption of THM precursors present in six U.S. waters was investigated by Amy and Chadik.[57] The raw water THMFP varied from 313 µg/L in the Mississippi River (LA) to 1087 µg/L in the Ilwaco Reservoir (WA). With optimum alum doses of 15 and 2.5 mg/L as Al, THMFP in the Mississippi River and Ilwaco Reservoir waters was reduced to 130 and 370 µg/L, respectively. The THMFP was then reduced to 81 and 196 µg/L, respectively, with the addition of 50 mg PAC per liter. Table 8 lists the results of Amy and Chadik[57] for reduction of THMFP in water from the Daytona Beach Aquifer (FL) and the Ilwaco

---

* Aqua Nuchar, Westvaco Corp., Covington, WV.

** Nuchar SA, Westvaco Corp., Covington, WV.

Table 5. Effect of PAC on Trihalomethane Formation Potential (THMFP) Concentration

Treatment Type	PAC Dose (mg/L)	Trihalomethanes (µg/L)				TTHM (µmol/L)	Reduction in TTHM (%)
		CHCl$_3$	CHBrCl$_2$	CHBr$_2$Cl	CHBr$_3$		
Settled water		NF*	NF	NF	NF	NF	
Settled Water + Cl$_2$		27	15.2	10.4	<0.1	0.37	0
Settled water + PAC − Cl$_2$	2	22	15.1	8.0	<0.1	0.31	16
	4	16.4	16.4	10.2	<1.0	0.36	2
	8	20	15.8	9.4	<1.0	0.32	14
	20	16	16.9	12.2	<1.0	0.29	22
	50	11	13.0	10.0	<1.0	0.22	41
	100	9	9.5	8.8	<1.0	0.18	51

† F—none found.

Table 6. Percentage Removals of THM Precursors with the Addition of PAC to New Orleans Water Compared with Those of an Agitated Control*

Treatment Sequence	THM Precursor Removal (%)		
	5 mg PAC/L	50 mg PAC/L	500 mg PAC/L
Mode 1			
PAC addition, coagulation-flocculation-sedimentation, filtration	0	31	84
Mode 2			
PAC addition, coagulation-flocculation-sedimentation, chlorination, filtration	8	41	77
Mode 3			
Coagulation-flocculation-sedimentation, PAC addition, chlorination, filtration	0	27	90

* Same treatment but with no PAC addition.

Table 7. Effect of PAC Dosage on the Removal of Chloroform Precursors from Ohio River Water*

PAC Dose (mg/L)	Chloroform Formation Potential	
	µg/L	Reduction (%)
0	236	0
10	215	9
25	170	28
50	127	46
100	85	64
1000	48	80

* 30°C, pH = 10.5—11.2, Cl_2 residual = 2—3 mg/L, stored 3 days.

Reservoir by using two types of PAC after coagulation. In the aquifer water, the value of THMFP to TOC for the raw water, 72.5 µg/mg, was virtually unchanged after coagulation. The ratio, however, dropped to an average of 50 µg/mg after treatment with 50 mg PAC per liter. This suggests that the portion of the natural organic matter that produces THMs is selectively removed by PAC. As for the Ilwaco Reservoir water, Table 8 shows that the THMFP to TOC value dropped from 135 µg/mg for the raw water to 108 µg/mg with the addition of 7.5 mg ferric chloride per liter, followed by a drop to 53 µg/mg after treatment with 50 mg PAC per liter. This also suggests the selective removal of THM-forming natural organics by PAC.

A review of these findings clearly suggests that the extent of THMFP reduction by PAC largely depends on the types of PAC and water being used. In the absence of a reliable surrogate parameter, the THMFP of a water sample needs to be measured directly. A standard method has been proposed for the determination of THMFP.[63]

Adsorption of THMs

Prechlorination is commonly practiced for the control of bacterial and algal growth in the treatment plant. In such a case, the THM formation reaction starts prior to the addition of PAC. By the time PAC is added, the water will contain some remaining THM precursors, THMs, and some residual chlorine. However, the oxidation-reduction reaction between chlorine and the surface of the activated carbon will result in the reduction of the adsorption capacity of activated carbon for phenolic compounds, as well as the destruction of the residual chlorine.[64,65] The same decrease in capacity may occur for other compounds. Therefore, the effect of PAC addition on the finished water THM concentration will not necessarily be due to the adsorption of the THM precursors or the THMs; the decrease in the residual chlorine concentration will result in lower THM formation,[56] which can be misinterpreted as higher THM removals by PAC. On the other hand, the destruction of the carbon surface will result in lower than expected removals of THMs per unit mass of PAC because of the reduction in carbon capacity.

Anderson et al.[66] investigated the control of THMs by PAC addition to the influent to the flocculation basin of a 40-mgd water treatment plant. In the first part of the study, chlorine was added to the raw water prior to the point of addition of PAC. In the second part, the chlorination point was moved to the flocculation basin effluent, allowing some contact between the water and the carbon in the absence of chlorine. With prechlorination, the THM concentration in the filter effluent was reduced from 40.2 to 27.5 µg/L with the addition of 7.3 mg PAC per liter, and from 37.4 to 23.8 µg/L with the addition of 21.6 mg PAC per liter. When the chlorine was added after the PAC, the THM concentration in the filter effluent dropped from 51.4 µg/L when no PAC was added to 28.2 µg/L with 19.4 mg PAC per liter added. The removals were higher than those predicted using the isotherms of Dobbs et al.[47] listed in Table 3. However, a good comparison cannot be made because chlorine was present during the adsorption of THMs.

In another full-scale study, Singley et al.[51] investigated the performance of PAC added to a softening plant for the control of THMs. Prechlorination of the raw water was also practiced during part of this study. The finished water THM concentration of 40 µg/L was virtually unchanged after the addition of 14.3 and 26.6 mg PAC per liter. The 3-day THMFP, however, decreased from 135 to 40 µg/L with the addition of the two carbon dosages, and the average TOC of the water was reduced form 14 to 10.4 mg/L (26.6% reduction). The THM results may have been affected by the removal of chlorine by the PAC. When the chlorination point was moved to the recarbonation units, the finished water THM concentration was still unchanged with the addition of 7.1 mg PAC per liter. However, the 3-day THMFP was reduced from 135 to 110 µg/L and the TOC was reduced from 12.6 to 4.4 mg/L (65.1% reduction). The much higher removals of TOC achieved when the chlorination point was moved (65.1%)

Table 8. Reduction of THMFP and TOC by Coagulation and Adsorption on 50 mg/L of Two Types of PAC in Water from Daytona Beach Aquifer, FL, and Ilwaco Reservoir, WA (After Amy and Chadik[57])

Coagulant Type	Coagulant Dose (mg/L)	After Coagulation			After Adsorption					
		THMFP (μg/L)	TOC (mg/L)	THMFP to TOC (μg/mg)	Carbon B			Carbon C		
					THMFP (μg/L)	TOC (mg/L)	THMFP to TOC (μg/mg)	THMFP (μg/L)	TOC (mg/L)	THMFP to TOC (μg/mg)
Daytona Beach Aquifer, FL										
Al	0	667	9.2	72.5	433	7.96	54.4	369	7.31	50.5
Al	10	509	6.88	74	297	6.13	48.4	309	5.77	53.6
Al	20	418	6.54	64	220	4.98	4.2	242	4.78	50.6
Al	30	383	4.5	69.8	218	4.22	51.7	265	4.18	63.4
Fe	15	548	7.65	71.6	345	6.21	55.6	321	5.97	53.8
Fe	30	401	6.76	59.3	257	5.41	47.5	328	5.09	64.4
Ilwaco Reservoir, WA										
Al	0	1,087	8.04	135.2	771	7.11	108.4	696	7	99.4
Al	1.25	579	5.34	108.4	569	5.64	100.9	442	5.14	86
Al	2.5	371	3.33	111.4	240	2.86	83.9	258	2.91	88.7
Fe	2.5	894	6.9	129.6	788	6.83	115.4	755	6.78	111.4
Fe	5	350	3.0	116.7	347	4.18	83.0	321	3.73	86.1
Fe	7.5	245	2.27	107.9	127	2.38	53.4	124	2.06	60.2

Table 9. Quantity of PAC Required as a Function of Dosage and System Size

System Size (mgd)	Quantity of PAC Required (lb/day)				
	5 mg PAC/L	10 mg PAC/L	25 mg PAC/L	50 mg PAC/L	75 mg PAC/L
0.1	2.9	5.84	14.6	29.2	43.8
0.5	14.6	29.2	73.0	146.0	219.0
1.0	29.2	58.4	146.0	292.0	438.1
2.5	73.0	146.0	365.0	730.0	1,095.2
5.0	146.0	292.0	730.0	1,460.3	2,190.4
10	292.0	584.1	1,460.3	2,920.5	4,380.8
15	438.1	876.1	2,190.4	4,380.8	6,571.2
25	730.1	1,460.3	3,650.6	7,301.3	10,951.9
50	1,460.3	2,920.5	7,301.3	14,602.6	21,903.9
75	2,190.4	4,380.8	10,951.9	21,903.9	32,855.9
100	2,920.5	5,841.0	14,602.6	29,205.2	43,807.8
150	4,380.8	8,761.5	21,903.9	43,807.8	65,711.8

compared with those achieved during prechlorination (26.6%) are consistent with a reduction in PAC capacity for organic compounds in the presence of free chlorine.

COST ANALYSIS OF PAC USE

The most important component of the cost of using PAC is the cost of the PAC itself. Over the last 30 years, the price of PAC has risen from about $0.07/lb to as high as $0.45/lb, an increase of more than 500%. During this time, the wholesale and producers' price indexes increased by about 200%. The largest increase in PAC costs occurred in the mid1970s following improvements in PAC products. In addition to the cost of the carbon, treatment costs for PAC usage* include capital and operation and maintenance requirements for process feed equipment.[67-70]** Preliminary cost estimates for PAC treatment have been developed for a range of system sizes and PAC dosage rates. Table 9 presents the quantity of PAC required (pounds per day) as a function of system size and dosage, assuming 70% utilization of plant capacity and continuous application of PAC. Design requirements for feed equipment were based on the maximum flow rate and PAC dosage rate. Tables 10 and 11 present annual cost estimates (dollars per year) for the purchase of PAC and process feed equipment, respectively.** Table 12 summarizes the unit cost estimates for PAC treatment (dollars per 1000 gal).

Figure 1 presents unit cost estimates for PAC treatment of ($/1000 gal) as a function of dosage rate for four system sizes. A significant decrease in cost occurs with increasing size for small systems up to about 1 mgd in capacity.

---

* All cost figures were updated to 1988.

** An interest rate of 10% and an equipment life span of 20 years were assumed for this analysis.

Table 10. Annual Cost Estimates for Purchase of PAC

System Size (mgd)	Cost of PAC ($/year)				
	5 mg PAC/L	10 mg PAC/L	25 mg PAC/L	50 mg PAC/L	75 mg PAC/L
0.1	426	853	2,132	4,263	6,395
0.5	2,132	4,263	10,658	21,316	31,974
1.0	4,263	8,526	21,316	42,632	63,963
2.5	10,658	21,316	53,290	106,580	159,899
5	21,316	42,632	106,580	213,204	319,798
10	42,632	85,279	213,204	426,393	639,597
15	63,963	127,911	319,798	639,597	959,395
25	106,595	213,204	532,988	1,065,989	1,598,977
50	213,204	426,393	1,065,989	2,131,979	3,179,969
75	319,798	639,597	1,598,977	3,197,969	4,796,961
100	426,393	852,786	2,131,979	4,263,959	6,395,938
150	639,597	1,279,179	3,197,969	6,395,938	9,593,923

Table 11. Annual Cost Estimates of PAC Feed Equipment

System Size (mgd)	Capital and O & M Costs of Feed Equipment ($/year)				
	5 mg PAC/L	10 mg PAC/L	25 mg PAC/L	50 mg PAC/L	75 mg PAC/L
0.1	2,125	2,135	2,165	2,365	2,595
0.5	2,420	2,620	3,155	4,065	4,665
1.0	3,270	3,655	4,715	5,925	6,980
2.5	5,095	6,005	7,820	11,130	14,180
5	7,555	8,765	12,680	18,810	23,480
10	11,215	13,620	21,260	29,095	36,990
15	14,080	18,290	27,740	38,800	48,290
25	20,295	26,425	37,580	53,315	67,865
50	32,105	39,940	58,995	86,115	113,065
75	43,230	54,290	80,000	119,520	152,320
100	48,980	61,795	95,155	138,685	174,905
150	68,495	85,885	133,725	186,525	249,135

Systems with capacities in the range of 1 to 100 mgd show only small gradual decreases in unit costs with increasing size, given a fixed PAC dosage. For example, assuming a PAC dose of 5 mg/L, the unit treatment cost drops from about $0.10 to $0.03/1000 gal with increasing system size from 0.1 to 1 mgd. By increasing system capacity from 1 to 100 mgd, the unit costs gradually decrease from $0.03 to $0.02/1000 gal. In this analysis, unit costs increase nearly linearly with respect to increasing dosage rate, given a fixed system size.

A typical maintenance dose of PAC for taste and odor control is approximately 5 mg/L. At this rate, quantities of PAC required for systems of 1.0- and 100-mgd size are about 1000 and 1,070,000 lb/year, respectively. Many systems must increase the PAC dosage rate for a period of time to treat seasonal and occasional variations in water quality. A utility may have to increase the PAC dose from 5 to as much as 75 mg/L for several weeks to reduce the concentrations

Table 12. Total Cost Estimates of PAC Treatment

| System Size (mgd) | PAC Treatment Cost Estimates ($/1000 gal) ||||||||||||||||
| | 5 mg PAC/L ||| 10 mg PAC/L ||| 25 mg PAC/L ||| 50 mg PAC/L ||| 75 mg PAC/L |||
	A*	B	C	A	B	C	A	B	C	A	B	C	A	B	C
0.1	0.083	0.017	0.100	0.083	0.033	0.116	0.085	0.083	0.168	0.092	0.167	0.259	0.101	0.250	0.351
0.5	0.019	0.017	0.036	0.021	0.033	0.054	0.025	0.083	0.108	0.032	0.167	0.199	0.036	0.250	0.286
1	0.013	0.017	0.030	0.014	0.033	0.047	0.018	0.083	0.101	0.023	0.167	0.190	0.027	0.250	0.277
2.5	0.008	0.017	0.025	0.009	0.033	0.042	0.012	0.083	0.095	0.017	0.167	0.184	0.022	0.250	0.272
5	0.006	0.017	0.023	0.007	0.033	0.040	0.010	0.083	0.093	0.015	0.167	0.182	0.018	0.250	0.268
10	0.004	0.017	0.021	0.005	0.033	0.038	0.008	0.083	0.091	0.011	0.167	0.178	0.014	0.250	0.264
15	0.004	0.017	0.021	0.005	0.033	0.038	0.007	0.083	0.090	0.010	0.167	0.177	0.013	0.250	0.263
25	0.003	0.017	0.020	0.004	0.033	0.037	0.006	0.083	0.089	0.008	0.167	0.175	0.011	0.250	0.261
50	0.003	0.017	0.020	0.003	0.033	0.036	0.005	0.083	0.088	0.007	0.167	0.174	0.009	0.250	0.259
75	0.002	0.017	0.019	0.003	0.033	0.036	0.004	0.083	0.087	0.006	0.167	0.173	0.008	0.250	0.258
100	0.002	0.017	0.019	0.002	0.033	0.035	0.004	0.083	0.087	0.005	0.167	0.172	0.007	0.250	0.257
150	0.002	0.017	0.019	0.002	0.033	0.035	0.003	0.083	0.086	0.005	0.167	0.172	0.006	0.250	0.256

* A — capital cost plus O & M feed equipment cost, B — PAC material cost only, C — total cost.

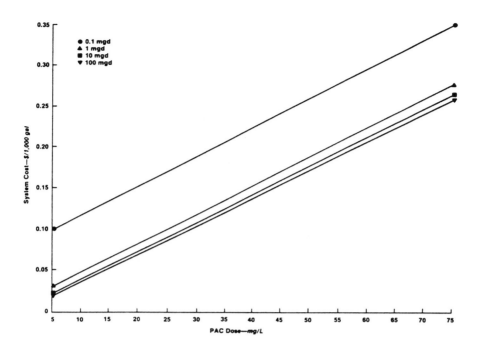

FIGURE 1. Preliminary cost estimates for PAC treatment as a function of dosage and system size.

of synthetic organic contamination resulting from an occasional chemical spill or seasonal agricultural runoff. For example, if a 10-mgd system increases its typical maintenance PAC dose rate from 5 to 50 mg/L for 4 weeks during the year, the additional PAC required is about 73,600 lb above the usual maintenance dosage. This results in a total PAC requirement of about 180,200 lb/year. This short-term but significant increase in the quantity of PAC required affects the treatment cost somewhat proportionally. The total cost estimate for the system, assuming only the 5-mg/L dose rate, is about $53,850/year. By including the 4-week increase in PAC dose to 50 mg/L, the total cost increases to about $84,660/year, an increase of about 60%. The reduction in the cost of applying PAC can be calculated in a similar manner if there are periods of time when no PAC is required.

SUMMARY AND CONCLUSIONS

Powdered activated carbon can be applied at different points in drinking water plants. In selecting the point of addition of PAC, consideration must be given to the degree of mixing and to minimizing short-circuiting, the adequacy of the time of contact between the PAC and the water, the PAC residence time, and the minimization of interference of treatment chemicals with the adsorption process. For example, good mixing can be achieved in the rapid mix and flocculation basin, but PAC in the sedimentation basin may not be effective because of poor mixing. When PAC is added to some treatment processes, such as floc-blanket reactors, significantly longer carbon residence times are achieved compared with conventional treatment. These longer residence times increase the removal of slowly adsorbing compounds and decrease the carbon usage rate.

The efficiencies of removal of taste- and odor-causing compounds by PAC vary, depending on the type and concentration of the compound causing the problem. High efficiencies of removal of compounds such as 2,4-dichlorophenol, geosmin, and 2-methylisoborneol were achieved with relatively low PAC doses (10 to 25 mg/L). Doses of PAC ranging from 75 to 620 mg/L, however, were required to remove *p*-nitrophenol and humic odor.

Powdered activated carbon has been effective for the removal of several biocides, depending on their type and concentration as well as the type of contactor. However, PAC doses as high as 475 mg/L were required to reduce the concentration of metolachlor from 50 to 1 μg/L.

The extent of adsorption of SOCs on PAC is strongly dependent on the type of compound being removed. In general, PAC has a relatively low capacity for VOCs compared with nonvolatile organics. It should be emphasized, however, that the means of application of PAC, as well as the contact time available, largely determine the percentage removal of these and other trace organics from drinking water.

The reported removals of THM precursors and THMs from drinking water by PAC range from poor to very good. The fraction of natural organic matter that forms THMs varies in different waters. Some studies have shown that the ratio of THMFP to TOC of a natural water decreases after treatment with PAC, indicating the preferential removal of THM precursors by PAC. However, depending on the type of water being treated, few studies have reported reduction in the TOC of the water with an insignificant drop in THM precursor concentration. Existing data indicate that no good correlation exists between the concentration of THM precursors and TOC in different types of water.

One of the main advantages of PAC is its low capital cost. Cost estimates of the addition of PAC at various dosages to different plant sizes are presented in this article. The annual cost of PAC decreases with increasing plant size. For a 10-mgd plant operating at full capacity and using a continuous PAC dose of 10 mg/L, the total cost is approximately $0.038/1000 gal, which amounts to $380/day or $138,700/year.

ACKNOWLEDGMENT

This work was supported by Lyonnaise des Eaux/Dumez, Paris, France; the AWWA Research Foundation; and the U.S. Environmental Protection Agency. The opinions expressed in this article are those of the authors and not necessarily those of the supporting agencies.

REFERENCES

1. AWWA Committee Report. "Measurement and Control of Organic Contaminants by Utilities," *J. AWWA* 69(5):267 (1977).
2. AWWA. 1984 Utility Operating Data, Denver, CO (1986).
3. Graese, S. L., Snoeyink, V. L., and Lee, R. G. "Granular Activated Carbon Filter-Adsorber Systems," *J. AWWA* 79(12):64 (1987).
4. Kruithof, J. C. et al. "Selection of Brands of Activated Carbon for Adsorptive Properties," in *Activated Carbon in Drinking Water Technology,* KIWA-AWWARF Rept. AWWARF, Denver, CO (1983).
5. Najm, I. N. et al. "Effect of Particle Size and Background Organics on the Adsorption Efficiency of PAC," *J. AWWA* 82(1):65 (1990).
6. AWWA. "AWWA Standard for Powdered Activated Carbon," B600-78, Denver, CO (1978).
7. Meijers, J. A. P. and van Der Leer, R. C. "The Use of Powdered Activated Carbon in Conventional and New Techniques," KIWA-AWWARF Rept. AWWARF, Denver, CO (1983).
8. Randtke, S. J. and Snoeyink, V. L. "Evaluating GAC Adsorption Capacity," *J. AWWA,* 75(8):406 (1983).
9. AWWA. *Water Quality and Treatment,* 3rd ed. (New York: McGraw-Hill, 1971).
10. Gauntlett, R. B. and Packham, R. F. "The Use of Powdered Activated Carbon in Water Treatment," Proceedings of Conference on Activated Carbon in Water Treatment, University of Reading, Water Research Association, Medmenham, England (April 1973).
11. Sontheimer, H. "The Use of Powdered Activated Carbon," in *Translation of Reports on Special Problems of Water Technology, Volume 9 — Adsorption,* EPA-600/9-76-030 (December 1976).
12. Najm, I. N. et al. "Powdered Activated Carbon in Floc-Blanket Reactors," 1989 AWWA Annual Conference, Los Angeles, CA (1989).
13. Sontheimer, H., Kölle, H., and Spindler, P. *Rohöl und Trinkwasser, Veröff. des Bereichs für Wasserchemie.* (Germany: University of Karlsruhe, 1967).
14. Dougharty, I. D. and Morris, R. L. "Studies on the Removal of Actinomycete Musty Tastes and Odors in Water Supplies," *J. AWWA,* 59(10):1320 (1967).
15. Gifford, J. S., George, D. B., and Adams, V. D. "Synergistic Effects of Potassium Permanganate and PAC in Direct Filtration Systems for THM Precursor Removal," *Water Res.* 23:10 (1989).

16. Kassam, K. et al. "Accumulation and Adsorption Capacity of Powdered Activated Carbon on a Slurry Recirculation Clarifier," 1989 AWWA Annual Conference, Los Angeles, CA (1989).
17. Hoehn, R. C. et al. "THM Precursor Control With Powdered Activated Carbon in a Pulsed-Bed, Solids Contact Clarifier," 1987 AWWA Annual Conference, Kansas City, MO (1987).
18. Richard, Y. Personal communication (1986).
19. Lettinga, G., Beverloo, W. A., and van Lier, W. C. "The Use of Flocculated Powdered Activated Carbon in Water Treatment," *Progr. Water Technol.* 10:537 (1978).
20. Haberer, K. and Normann, S. Untersuchungen zu einer Neuartigen Pulverkohle-Filtrationstechnik für die Wasseraufbereitung, *Vom Wasser* 49:331 (1979).
21. Haberer, K. and Normann, S. "Entwicklung eines Kurztalt-Filtrations Verfahrens zum Einsatz vor Pulverkohle in der Wasseraufbereitung," *Gas Wasserfach Wasser Abwasser,* 118:393 (1980).
22. Hoehn, R. C. et al. "A Pilot-Scale Evaluation of the Roberts-Haberer Process for Removing Trihalomethane Precursors From Surface Water With Activated Carbon," 1984 AWWA Annual Conference, Dallas, TX.
23. Sontheimer, H., Crittenden, J. C., and Summers, R. G. *Activated Carbon for Water Treatment,* 2nd ed. (Germany: DVGW-Forschungsstelle, Engler-Bunte-Institut, Universität Karlsruhe, (1988).
24. Snoeyink, V. L. and Suidan, M. T. "Dechlorination by Activated Carbon and Other Reducing Agents," *Disinfection: Water and Wastewater,* J. D. Johnson, Ed. (Ann Arbor, MI: Ann Arbor Science, 1975).
25. McGuire, M. J. and Suffet, I. H. "Aqueous Chlorine/Activated Carbon Interaction." *J. Environ. Eng. Div. ASCE* 119(3):3629 (1984).
26. Lalezary-Craig, S. et al. "Optimizing the Removal of Geosmin and 2-Methylisoborneol by Powdered Activated Carbon," *AWWA,* 80(3):73 (1988).
27. Lalezary, S., Pirbazari, M., and McGuire, M. J. "Evaluating Activated Carbons for Removing Low Concentrations of Taste-and Odor-Producing Organics," *J. AWWA* 78(11):76 (1986).
28. Miltner, R. J. et al. "Final Internal Report on Carbon Use Rate Data," U.S. EPA, Cincinnati, OH (June 1987).
29. Frick, B. "Adsorptionsgleichgewichte zwischen Aktivkohle und organischen Wasserinhaltsstoffen in Mehrstoffgemischen bekannter und unbekannter Zusammensetzung," PhD Dissertation, University of Karlsruhe, Germany (1980).
30. Hansen, R. E. "The Costs of Meeting the New Water Quality Standards for Total Organics and Pesticides." 1975 AWWA Annual Conference, Minneapolis, MI (1975).
31. Burtschell, R. H. et al. "Chlorine Derivatives of Phenol Causing Taste and Odor," *J. AWWA* 51(2):205 (1959).
32. Sigworth, E. A. "Identification and Removal of Herbicides and Pesticides, *J. AWWA* 57(8):1016 (1965).
33. Westvaco Corp. "Taste and Odor Control in Water Purification," Covington, WV (1970).
34. Brown, G. N., Hyndshaw, D. B., and van Dongen, D. B. "Powdered Activated Carbon Removal of Tastes and Odors Plus," *Proc. AWWA Ann. Conf.* 97:2 (1977).

35. Aly, O. M. and Faust, S. D. "Removal of 2,4-Dichlorophenoxy-acetic Acid Derivatives From Natural Waters," *J. AWWA,* 51(2):221 (1965).
36. Singley, J. E. and Ervin, A. L. "East Drive Well Field Powdered Activated Carbon Study," Report of the City of North Miami Beach, FL (1978).
37. Symons, J. M. et al. "Interim Treatment Guide for Controlling Organic Contaminants in Drinking Water Using Granular Activated Carbon," Water Supply Research Division, U.S. EPA, Cincinnati, OH (Jan. 1978).
38. Miltner, R. J. et al. "Treatment of Seasonal Pesticides in Surface Waters," *J. AWWA* 81(1):43 (1989).
39. *Fed. Reg.* 43:28 (1978).
40. U.S. EPA. "National Primary and Secondary Drinking Water Regulations; Proposed Rule," 40 CFR Parts 141, 142, 143, Vol. 54:97:22062, Cincinnati, OH (May 1989).
41. Robeck, G. G. et al. Effectiveness of Water Treatment Processes in Pesticide Removal. *J. AWWA,* 57(2):181 (1965).
42. Richard, J. J. et al. "Analysis of Various Iowa Waters for Selected Pesticides: Atrazine, DDE, and Dieldrin — 1974," *Pestic. Monit. J.* 9:3 (1975).
43. U.S. EPA. "Powdered Activated Carbon Treatment for Removal of Trace Organic Compounds From the New Orleans Water Supply," U.S. EPA Grant R-804404 (March 1976).
44. DiFilippo, J. D., Copeland, L. G., and Peil, K. M. "Evaluation of Powdered Activated Carbon for the Removal of Trace Organics at New Orleans, LA. EPA-600/2-81-027, U.S. EPA, Cincinnati, OH (1981).
45. *Fed. Reg.* 52:130 (1987).
46. *Fed. Reg.* 54:97 (1989).
47. Dobbs, R. A., Middendorf, R. J., and Cohen, J. M. "Carbon Adsorption Isotherms for Toxic Organics," MERL, U.S. EPA, Cincinnati, OH (May 1978).
48. Seeger, D. R., Slocum, C. J., and Stevens, A. A. "GC-MS Analysis of Purgeable Contaminants in Source and Finished Drinking Water," Annual Conference on Mass Spectrometry and Applied Topics, St. Louis, MO, May 28-June 2, 1978.
49. Rexing, M. Personal communication (1978).
50. Singley, J. E. et al. "Minimizing Trihalomethane Formation in a Softening Plant," MERL, U.S. EPA, Cincinnati, OH (1977).
51. Singley, J. E., Beaudet, B. A., and Ervin, A. L. "Use of Powdered Activated Carbon for Removal of Specific Organic Compounds," in *Proceedings on Seminar on Controlling Organics in Drinking Water,* 1979 AWWA Annual Conference, San Francisco, CA (1979).
52. Rook, J. J. "Formation of Haloforms During Chlorination of Natural Waters," *Proc. Water Treat. Exam* 23(2):234 (1974).
53. Bellar, T. A., Lichtenberg, J. J., and Kroner, R. C. "The Occurrence of Organohalides in Chlorinated Drinking Water, *J. AWWA,* 66(12):703 (1974).
54. *Fed. Reg.* 44:231 (1979).
55. Lange, A. L. and Kawczynski, E. "Trihalomethane Studies, Contra Costa County Water District Experience," Proceedings Water Treatment Forum VII, California-Nevada AWWA Section, Palo Alto, CA., April 1978.

56. Hentz, L. H., Jr., Hoehn, R. C., and Randall, C. W. "Haloform Formation in Water From Peat-Derived Humic Substances and the Reducing Effects of Powdered Activated Carbon and Aluminum," *Proceedings* 1980 AWWA Annual Conference, Atlanta, GA (1980).
57. Amy, G. L. and Chadik, P. A. "PAC and Polymer Aided Coagulation for THM Control," in *Proceedings National Conference on Environmental Engineering* (ASCE, July 1982).
58. Hoehn, R. C. et al. "Chlorination and Water Treatment for Minimizing Trihalomethanes in Drinking Water," in *Proceedings of Conference on Water Chlorination: Environmental Impact and Health Effects,* Vol. 2, R. H. Jolley, H. Gorchev, and D. H. Hamilton, Eds., Ann Arbor, MI (1977).
59. Love, O. T. et al. "Treatment for the Prevention or Removal of Trihalomethanes in Drinking Water, DWRD, U.S. EPA, Cincinnati, OH (1976).
60. Wood, P. R. and DeMarco, J. "Effectiveness of Various Adsorbents in Removing Organic Compounds From Water," 176th ACS Mtg.: Activated Carbon Adsorption of Organics From the Aqueous Phase, Miami Beach, FL, September 10-15, 1978.
61. Symons, J. M. et al. "Interim Treatment Guide for the Control of Chloroform and Other Trihalomethanes," DWRD, U.S. EPA, Cincinnati, OH (June 1976).
62. Zogorski, J. S., Allgeier, G. D., and Mullins, R. L., Jr. "Removal of Chloroform from Drinking Water," *Res. Rept. 111.* University of Kentucky Water Resources Research Institute, Lexington, KY (June 1978).
63. *Standard Methods for the Examination of Water and Wastewater,* 17th ed. (Washington, D.C.: HPHA, AWWA, and WPCF, 1989).
64. Suidan, M. T., Snoeyink, V. L., and Schmitz, R. A. "Performance Predictions for the Removal of Aqueous Free Chlorine by Packed Beds of Granular Activated Carbon," *AIChE Symp. Ser.,* 73:18 (1976).
65. Suidan, M. T., Snoeyink. V. L., and Schmitz, R. A. "Reduction of Aqueous HOCl With Granular Activated Carbon," *J. Environ. Eng. Div. ASCE* 103:677 (1977).
66. Anderson, M. C. et al. "Control of Trihalomethanes Using Powdered Carbon," *Natl. Conf. Environ. Eng. ASCE* (July 1980).
67. Lee, R. Personal communication. (November 1988).
68. Gumerman, R. C., Russell, L., and Hansen, S. P. "Estimating Water Treatment Cost, Vol. II, Cost Curves Applicable to 1 to 200 mgd Treatment Plants," EPA-600/2/79-162b (1979).
69. Gumerman, R. C., Burris, B. E., and Hansen, S. P. "Estimation of Small System Water Treatment Costs," U.S. EPA 600/S2-84-184 (March 1985).
70. Fisher, J. L. Personal communication (November 1988).
71. Dobbs, R. A. and Cohen, J. M. "Carbon Adsorption Isotherms for Toxic Organics," EPA-600/8-80-023 (April 1980).
72. El-Dib, M. A., Moursy, A. S., and Badaway, M. I. "Role of Adsorbents in the Removal of Soluble Aromatic Hydrocarbons From Drinking Water," *Water Res.* 12:1131 (1978).
73. Cohen, J. M. et al. "Effects of Fish Poison on Water Supplies. I. Removal of Toxic Materials," *J. AWWA* 52(12):1551 (1960).
74. Steiner, J. IV, Singley, J. E., and Edward J. "Methoxychlor Removal From Potable Water," *J. AWWA* 71(5):284 (1979).

CHAPTER 4

Development of a Method to Predict the Adsorption of Organic Chemicals on Activated Carbon

D. J. W. Blum and I. H. Suffet

Contamination of water supplies by organic chemicals is a key problem confronting the water supply industry. There are a variety of techniques available for removal of organic pollutants. Prominent among these techniques is activated carbon adsorption. However, when confronted with a contamination problem, particularly an urgent one, adsorption isotherms may not be available for the chemicals involved. Thus it would be beneficial in these instances to have a method by which adsorbability could be predicted in the absence of experimental data. This method could be incorporated into an "expert system" capable of predicting the potential success of activated carbon adsorption and comparing it with other water treatment options. A research objective of developing such an expert system for water treatment processes including activated carbon adsorption is under investigation by Lyonnaise Des Eaux, Le Pecq, France. The intent of the present research is to develop a method to predict the adsorbability of organic chemicals that have no determined isotherm.

Structural activity relationships (SARs) are used here to estimate adsorbability of organic chemicals in the absence of isotherms. SARs have been developed previously for prediction of two phase equilibrium systems such as air-water (Henry's Law constant),[1] liquid-liquid extraction methods,[2] and octanol-water partition coefficients.[3]

The first step of SAR development is to compile a Freundlich isotherm database that can be used as a training set for prediction of adsorption of organic chemicals on activated carbon. Then several quantitative SAR approaches are evaluated. The quantitative SAR methods studied include linear solvation energy relationships of Kamlet,[4] molecular connectivity,[3] and correlations employing octanol-water partition coefficients or parameters based on the Polyani potential theory.[5]

In the absence of experimentally determined isotherm data for a chemical, values predicted by an SAR can be used to evaluate the potential for activated carbon adsorption. As an initial study to develop such an SAR for the adsorption of organic chemicals on activated carbon, this research completes the following tasks:

1. Identification of data to be used as a training set for the development of a SAR.
2. Comparison of the potential success of four SAR methods using a relatively small initial database.
3. Development of a SAR using the optimum method with an expanded database.

DATA

There are a few important criteria to consider in identifying data, the training set, for use in the development of a SAR for activated carbon adsorption. A SAR is only applicable to the type of compounds used in its development. Therefore, the data used should cover a wide range of chemical structures so that one can determine how generally applicable the SAR is. The type of compounds for which one anticipates using the SAR for prediction should be well represented in the training set. The testing conditions under which the data were collected should be applicable to the conditions for which one would like to predict adsorption. Data considered here were for F 300 carbon, ambient temperatures, and pH between 5 and 9. Finally, it is helpful in the initial development of an SAR to use data collected by a single method or source so that variability due to experimental method is minimized.

In order to develop an SAR for activated carbon adsorption it is desirable to characterize the attraction of the chemical to the carbon with a single numerical descriptor. The Freundlich isotherm uses two parameters, K and n to model adsorption:

$$X = K(C^{1/n}) \tag{1}$$

where X = mg of adsorbate adsorbed per g of carbon

C = mg of adsorbate per L of solution in equilibrium

K = constant in units of L per g

n = dimensionless constant

Different approaches have been used to define the relationship between adsorbate and carbon with one descriptor. All of these methods are supported by the observation that at low concentrations, the quantity adsorbed (X) varies linearly with solution equilibrium concentration (C).[6] Arbuckle[7] used K at one constant value of C. Abe et al.[8] used:

$$\alpha = \lim_{c \to 0} (X/C) \quad (2)$$

This function is based on the observation that as C becomes small, X/C tends to become constant, that is, the isotherm is linear (n = 1) at low concentrations. Abe obtained α by extrapolating adsorption data back to the origin. Belfort et al.[9] used the Langmuir model rather than the Freundlich. However, also based on the assumption that the isotherm is linear at low concentrations, they used the slope of the isotherm at infinite dilution, $Q°b$, as a single descriptor.

A similar approach was also adopted here: X/C was assumed to be a constant at low concentrations. $(X/C)_{min}$ was used (in liters per grams) as a descriptor of adsorbate per carbon affinity. The question then becomes how to use available databases of isotherm data to determine $(X/C)_{min}$. Most isotherm data are presented with K, n, and the range of concentrations over which the parameters were determined. When the data are presented in this manner, it is possible to use the K and n values to determine X at the lowest value of C tested (C_{min}), in order to obtain $(X/C)_{min}$. There are two limitations to this approach. One, it is not known whether C_{min} is actually in the linear portion of the isotherm. Two, K and n, are used to back-calculate a value of X at C_{min} and this is less accurate than using the actual X data at C_{min}. However, if the only data available are the Freundlich parameters and concentration range, this is the only way to obtain an estimate of $(X/C)_{min}$. Other approaches are possible depending on how the isotherm data are presented. Our interpretation of the data as presented in several sources is described here.

A number of databases were considered for inclusion in this study. Alpha values from Abe et al.[8] and Giusti et al.[10] were used directly as $(X/C)_{min}$. These two sources contain a total of 38 aliphatic compounds including ketones, esters, aldehydes, ethers, and alcohols. Theoretically, Belfort[9] $Q°b$ data should be equivalent to $(X/C)_{min}$, but these data were found to correspond poorly to Abe and Giusti data. Arbuckle[7] presents isotherm parameters and therefore the approach described earlier was used to estimate $(X/C)_{min}$. There is very good correspondence between Abe and Giusti, and Arbuckle data when interpreted in this manner. However, Arbuckle tested alcohols, aldehydes, esters, ketones, and ethers — classes also tested by Abe and Giusti — and therefore these data add little to the diversity of the database.

Two sources, Sontheimer et al.[11] and Suffet et al.[12] provide compilations of isotherm data from many sources. These include data for important environmental pollutants including halogenated compounds. Each lists K, n, and concentration ranges and therefore $(X/C)_{min}$ can be estimated from the data. Much of the data in both sources is from Dobbs and Cohen.[13] Dobbs and Cohen data were consulted directly. Their published data contain all of the raw isotherm data and thus $(X/C)_{min}$ can be calculated for the lowest concentration they tested (eliminating the need to back-calculate X). In addition the assumption of linearity can be verified by looking at a few of their data points at low concentrations.

Initially, data from Abe et al.,[8] Giusti et al.,[10] and Dobbs and Cohen[13] were used for the following reasons: these sources provide data on a wide range of chemical classes including simple aliphatic chemicals (mostly from Abe and Giusti) and important toxic pollutants (mostly from Dobbs). The data are presented in such a way that $(X/C)_{min}$ can be ascertained easily. Finally, the use of a limited number of data sources reduces the variability due to experimental differences. After a SAR is developed, additional data, particularly from Suffet et al.[12] and Sontheimer et al.,[11] can be added to expand the range of applicability of the equation and improve its accuracy.

COMPARISON OF FOUR SAR METHODS

Once a database of chemicals is identified to serve as a training set for the development of a SAR, a group of parameters describing the chemicals must be identified which are expected to relate to the property of interest — the adsorption of the chemical to activated carbon. These parameters should be simple descriptors based on chemical structure or basic empirical descriptors that are easy to obtain experimentally. Statistical techniques, here multiple linear regression, are used to relate the parameters to the chemical property. The final equation developed by multiple linear regression then allows the adsorption of an untested chemical to be estimated from the parameters. Four established SAR methods, which are really groups of chemical-describing parameters, were used initially with a small database to identify the most promising method or methods. The methods were:

1. Linear solvation energy relationships (LSER)
2. Polyani potential theory
3. Octanol-water partitioning (Log P)
4. Molecular connectivity

Two simple statistics were used to compare the accuracy of the SARs: r^2, the coefficient of determination, and s, the root mean square error.

The initial database for SAR development consisted of approximately 40 compounds. $(X/C)_{min}$ were taken from Abe, Giusti, and Dobbs as described previously. The compounds cover a range of chemical classes. They were selected for the initial data set because their parameters for all four SAR methods were

Table 1. Comparison of Success of Four SAR Methods

	Aliphatic Compounds			Aromatic Compounds		
QSAR method	n	r^2	s	n	r^2	s
LSER	23	0.95	0.30	16	0.89	0.35
Polyani	14	0.79	0.48	8*	0.82	0.32
Log P	17	0.90	0.40	17	0.63	0.52
Molecular connectivity						
1 index	18	0.63	0.74	14	0.67	0.61
2 indices	18	0.76	0.61	14	0.78	0.52
3 indices	18	0.85	0.52	14	0.91	0.36

* Phenols excluded.

readily available. We found initially that aliphatic and aromatic compounds did not fit into the same correlations. Therefore, separate regressions were run for aliphatic and aromatic chemicals.

Table 1 contains a summary of the results. Each of the four SAR methods are described briefly followed by an assessment of the potential of each method for a SAR for activated carbon adsorption.

Linear solvation energy relationships developed by Kamlet and co-workers have been successful at correlating many diverse chemical properties, including adsorption on activated carbon, which depend on solute-solvent interactions. They are based on four molecular characteristics called solvatochromic parameters:

1. V_i, the intrinsic molecular volume
2. π^*, a measure of polarity or polarizability
3. β_m, a measure of the ability to participate in hydrogen bonding as a hydrogen bond accepter
4. α_m, a measure of the ability to participate in hydrogen bonding as a hydrogen bond donor

The LSER method resulted in the most accurate SARs for the initial database. Because of its accuracy and because it is based on parameters with easily interpreted meanings, it is a good exploratory method for developing accurate SARs, identifying outlying data or classes of data, and providing an indication of the solubility properties related to activated carbon adsorption. However, it may be wise to use an additional method in conjunction with LSER because the database of LSER parameters is limited. LSER parameters are available or can be estimated for approximately 1000 chemicals.

Polyani potential theory[5] has been used to relate the loading onto activated carbon to the function:

$$T/V * \log(C_s/C) \qquad (3)$$

where T = absolute temperature

V = molar volume, cm³/mol

C_s = aqueous solubility

C = concentration in solution at equilibrium

An adjustment factor is applied by dividing this function by:

$$X/0.236 - 0.28$$

where X = $n_D^2 - 1/n_D^2 + 2$

n_D = refractive index

In order to predict adsorption from the theory, an additional adjustment is needed, the "adsorbate volume adjustment." This adjustment is empirical and thus the theory does not yet allow prediction of isotherm data directly from the physical and chemical descriptors. However, the function shown here, called the Polyani parameter, can be used in correlations to predict adsorption. This was done by Arbuckle[7] to correlate adsorption for alcohols, aldehydes, and ketones.

As seen in Table 1, the Polyani method had reasonable success with r^2 of 0.80 for aliphatic chemicals. For aromatic chemicals, benzenes and phenols were not successfully included in the same correlation. One limitation of the method is that it depends on aqueous solubility. Experimental aqueous solubility data are often difficult to find and erratic. Miscible liquids pose a problem. In the correlations done here, experimental data found in Syracuse Research Corporation on-line database[18] were used. It may, in fact, improve the accuracy of the Polyani method to use solubility calculated by a SAR such as those developed by Kamlet et al.[14] or Nirmalakhandan and Speece.[17] In this way the solubility data would be internally consistent. The Polyani method is satisfying to use in that it is based on a theory of adsorption and would presumably improve as the theoretical understanding of adsorption improved.

Octanol-water partition coefficients measured as Log P have been used very extensively in SARs for toxicity where it models the relative partitioning between the aqueous phase and the more nonpolar lipidlike biophase. It may also be employed to model the partitioning between the polar aqueous phase and the nonpolar carbon in adsorption. Log P values can be determined experimentally with fair ease and accuracy. They can also be estimated by a computer-based expert system, CLOGP3.[15] As seen in Table 1, Log P had fair success here, particularly for the aliphatic compounds.

Molecular connectivity is a method of describing molecular structure based solely on the molecule's bonding and branching patterns as opposed to physical or chemical characteristics. It has been developed and used in many successful SARs by Kier and Hall.[3] Using a simple algorithm, a series of indices (called zero order, first order, etc.) describing increasingly larger molecular fragments is computed for chemicals based solely on their structural formulas. Correlations are found by selecting indices which explain the greatest amount of variance in the adsorption data and then finding coefficients to optimize the correlation statistically.

Table 1 lists the statistics for the molecular connectivity SARs. The method appeared promising, with greater accuracy achieved as more indices were included in the correlations. (Of course, if developed further, care must be taken that the index-to-data ratio is not too high to diminish statistical significance.) This method holds the distinct advantage of allowing the parameters to be computed by structural rules (nonempirically) with complete accuracy. It is less satisfying from a scientific standpoint than other methods, particularly as an initial exploratory technique, because the indices do not refer to specific known physical or chemical parameters. From a pragmatic viewpoint, however, adsorption of any chemical can be estimated without reliance on any experimental data.

LSER CORRELATION FOR EXPANDED DATABASE

The intent of the initial correlations was to identify the most promising SAR method or methods for further development. Expanding the database on which the correlations are based provides a more generally applicable and statistically valid SAR. The database was expanded to include 80 compounds (See Table 2). LSER was used for the correlation because of its success with the initial database and its suitability noted earlier for exploratory analyses. The following correlations were obtained:

Aliphatic compounds — LSER

$$\log(X/C)_{min} = -0.93(+/-0.36) + 5.88$$
$$(+/-0.66)Vi/100 - 4.99(+/-0.54)\beta_m$$

where $n = 50$
$r^2 = 0.91$
$s = 0.33$

Table 2. LSER SAR

	Chemical	Abe[8]	Guisti[10]	Dobbs[13]	Predict	Residual
1	Chlorobenzene			1.94	2.24	−0.29
2	1,2-Dichlorobenzene			2.21	2.55	−0.34
3	1,3-Dichlorobenzene			2.00	2.47	−0.47
4	1,4-Dichlorobenzene			2.39	2.38	0.01
5	1,2,4-Trichlorobenzene			2.28	2.63	−0.35
6	Hexachlorobenzene*			4.33	3.04	1.30
7	Nitrobenzene	0.91		3.57	2.83	0.44
8	2,6-Dinitrotoluene			2.70	3.28	−0.58
9	Phenol			1.25	1.54	−0.30
10	2-Chlorophenol			1.65	1.79	−0.14
11	2,4-Dichlorophenol*			3.91	1.78	2.13
12	2,4,6-Trichlorophenol			2.21	1.93	0.27
13	4-Nitrophenol			2.45	2.19	0.26
14	2,4-Dimethylphenol			2.19	1.88	0.30
15	4-Nitroaniline			3.69	2.87	0.82
16	p-Xylene			1.74	2.07	−0.32
17	Acetophenone			2.13	2.72	−0.59
18	4-Aminobiphenyl			3.69	3.83	−0.14
19	1-Naphthylamine			3.10	2.71	0.38
20	2-Naphthylamine			2.88	2.71	0.17
21	1-Naphthol			2.16	2.19	−0.03
22	2-Naphthol			2.39	2.35	0.04
23	Benzoic acid			0.60	1.64	−1.04
24	Anthracene			3.76	3.17	0.58
25	Benzo(a)pyrene			4.04	4.24	−0.20
26	Benzo(ghi)perylene			4.02	4.64	−0.61
27	Fluorene*			4.38	2.84	1.54
28	Fluoranthene			3.96	3.38	0.58
29	Benzo(b)fluoranthene			4.56	4.02	0.53
30	Benzo(k)fluoranthene			4.27	4.02	0.24
31	Methanol		−1.92		−1.82	−0.10
32	Ethanol		−1.35		−1.38	0.04
33	1-Propanol	−0.74	−0.96		−0.81	−0.02
34	1-Butanol	−0.17	−0.12		−0.24	0.10
35	1-Pentanol	0.35	0.30		0.31	0.01
36	1-Hexanol	0.96			0.88	0.08
37	Methylene chloride			−0.13	0.55	−0.68
38	1,2-Dichloroethane		0.57	0.93	1.17	−0.38
39	1,1-Dichloroethane			0.96	1.17	−0.21
40	1,1,1-Trichloroethane			2.02	1.62	0.40
41	1,1,2-Trichloroethane			1.63	1.62	0.01
42	1,1,2,2-Tetrachloroethane			2.49	2.20	0.29
43	Hexachloroethane			3.70	3.22	0.49
44	1,2-Dichloropropane		1.16	1.40	1.75	−0.46
45	1,1-Dichloroethylene			1.60	1.21	0.39
46	trans-1,2-Dichloroethylene			1.64	1.21	0.44
47	Trichloroethylene			2.22	1.71	0.51
48	Tetrachloroethylene			2.69	2.22	0.47
49	Acetone	−0.90	−0.86		−1.09	0.21
50	Ethyl ether	−0.26			−0.31	0.04
51	Isopropyl ether		0.60		0.83	−0.23
52	Chloroform*			−0.15	1.08	−1.24
53	Carbon tetrachloride			1.33	1.59	−0.26
54	2-Butanone	−0.44	−0.26		−0.52	0.18
55	2-Hexanone	0.71	0.55		0.61	0.02

Table 2 (continued). LSER SAR

	Chemical	Abe[8]	Guisti[10]	Dobbs[13]	Predict	Residual
56	2-Pentanone	0.14	0.24		0.05	0.14
57	4-Methyl-2-pentanone		0.71		0.62	0.09
58	5-Methyl-2-hexanone		0.72		1.18	−0.46
59	Cyclohexanone		0.17	0.84	0.11	0.51
60	Methyl acetate	−0.56	−0.74		−0.53	−0.11
61	Ethyl acetate	−0.04	−0.18		−0.11	0.01
62	n-Propyl acetate	0.62	0.39		0.48	0.03
63	n-Butyl acetate	1.20	0.70		1.03	−0.02
64	n-Amyl acetate		0.84		1.60	−0.76
65	Isopropyl acetate		0.20		0.48	−0.28
66	Isobutyl acetate		0.60		1.03	−0.43
67	di-n-Propylamine		0.54		−0.14	0.67
68	di-n-Butylamine		0.80		1.00	−0.20
69	Propionaldehyde	−0.82	−0.70		−0.74	−0.02
70	Butyraldehyde	−0.19	−0.14		−0.15	−0.01
71	Valeraldehyde	0.38			0.42	−0.03
72	Acrolein		−0.63		−0.97	−0.34
73	Vinyl acetate		0.11		0.10	0.01
74	di-n-Propyl ether	0.84			0.88	−0.05
75	2-Propanol		−1.21		−1.12	−0.09
76	2-Methyl-1-propanol		−0.37		−0.30	−0.07
77	2-Methyl-2-propanol		−0.65		−0.85	0.19
78	2-Ethyl-1-butanol		0.73		0.82	−0.09
79	2-Ethyl-1-hexanol		2.03		2.02	0.01
80	2-Propen-1-ol		−0.89		−0.68	−0.20
81	Ethyl chloride			−0.20	0.64	−0.84

Note: All data in log $(X/C)_{min}$ L/g.

* Chemicals omitted from regression.

Chloroform was an outlier and was omitted from the regression. π^* and α_m were not significant parameters.

Aromatic compounds — LSER

$$\log(X/C)_{min} = 1.81(+/-0.76)Vi/100 + 1.67(+/-0.82)\pi^* - 1.03(+/-0.64)\alpha_m$$

where n = 27

r² = 0.80

s = 0.48

Three outliers were excluded from this equation: 2,4-dichlorophenol, fluorene, and hexachlorobenzene. The omission of these outliers did not significantly change the equation coefficients, but did result in improved fit. Their lack of agreement probably reflects experimental variability rather than any limitation

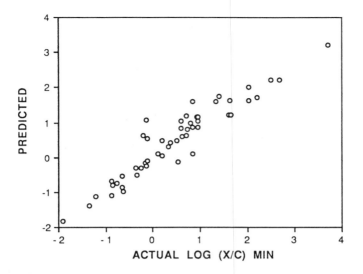

FIGURE 1. LSER SAR for aliphatic compounds.

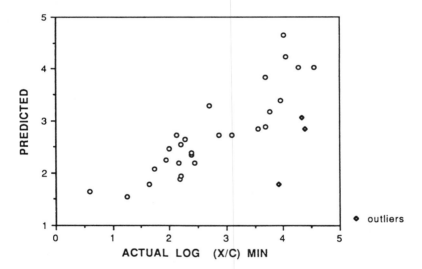

FIGURE 2. LSER SAR for aromatic compounds.

of the SAR method because they appear to be unrelated chemically. β_m was not a significant parameter.

The aliphatic and aromatic equations are significantly different. Table 2 shows the data included in the regression and the predicted values. Figures 1 and 2 graph each of the regressions.

CONCLUSIONS AND RECOMMENDATIONS FOR FUTURE WORK

This study has compared four potential SAR methods for the estimation of chemical adsorption on activated carbon. The most accurate method, LSER, was then used with a database of approximately 80 compounds to develop a preliminary SAR able to estimate adsorption with an accuracy of better than a half order of magnitude. Separate equations were needed for aliphatic and aromatic chemicals. The reason for this should be investigated further.

These preliminary SAR equations can be used in an expert system for estimating adsorption of untested chemicals provided that the chemicals are similar to those in the training set (Table 2) and that LSER parameters are available for the untested chemicals. Additional data will be incorporated into the LSER correlations to further improve its accuracy and extend its range of applicability.

An additional SAR method which will not have the limitation of the finite LSER parameter database may be desired in the expert system to supplement the LSER SAR. Although the method based on the Polyani theory is attractive for its theoretical base, molecular connectivity appears to be more accurate, has no limitations on the types of compounds that can be included, and its parameters are easily calculated. A SAR using molecular connectivity will be developed.

ACKNOWLEDGMENT

This work was supported by Lyonnaise Des Eaux/Dumez, Le Pecq, France. Project officers were Drs. C. Anselme and J. Mallevialle.

REFERENCES

1. Nirmalakhandan, N. and Speece, R. E. "Prediction of Henry's Constant Based on Molecular Structure," *Environ. Sci. Technol.* 22:1349 (1988).
2. Magnussen, T. et al. *Fluid Phase Equilibria* 4:151 (1980).
3. Kier, L. B., and Hall, L. H. *Molecular Connectivity in Chemistry and Drug Research,* (New York: Research Studies Press Ltd., John Wiley & Sons Inc., 1976).
4. Kamlet, M. J., Doherty, R. M., Abboud, J. M., Abraham, M. H., and Taft, R. W. "Solubility, A New Look," *Chem. Technol.* (1986).
5. Manes, M. "The Polanyi Adsorption Potential Theory and its Applications to Adsorption from Water Solution onto Activated Carbon," in *Activated Carbon Adsorption of Organics from the Aqueous Phase,* Suffet, I. H., and McGuire, M. J., Eds. (Ann Arbor, MI: Ann Arbor Science, 1980).
6. McGuire, M. J. and Suffet, I. H., "The Calculated Net Adsorption Energy Concept," in *Activated Carbon Adsorption of Organics from the Aqueous Phase,* (Ann Arbor, MI: Ann Arbor Science, 1980).

7. Arbuckle, W. B. "Estimating Equilibrium Adsorption of Organic Compounds on Activated Carbon from Aqueous Solution," *Environ. Sci. Technol.* 15:812-819 (1981).
8. Abe, I., Kamaya, H., and Ueda, I. "Activated Carbon as a Biological Model: Comparison between Activated Carbon Adsorption and Oil-Water Partition Coefficient for Drug Activity Correlation," *J. Pharm. Sci.* 77:166-168 (1988).
9. Belfort, G., Altshuler, G. L., Thallam, K. K., Feerick, C. P., Jr., and Woodfield, K. L. "Selective Adsorption of Organic Homologues onto Activated Carbon from Dilute Aqueous Solutions: Solvophobic Interaction Approach. IV. Effect of Simple Structural Modifications with Aliphatics, *AICHE J.* 30:197-207 (1984).
10. Kamlet, M. J., Doherty, R. M., Abraham, M. H., and Taft, R. W. "Linear Solvation Energy Relationships. XXXIII. An Analysis of the Factors that Influence Adsorption of Organic Compounds on Activated Carbon," *Carbon* 23:549-554 (1985).
11. Sontheimer, H., Crittenden, J. C., and Summers, R. S. *Activated Carbon for Water Treatment,* AWWA Research Foundation (1988).
12. Suffet, I. H. "An Evaluation of Activated Carbon for Drinking Water Treatment: National Academy of Science Report," *J. Am. Water Works Assoc.* (1980).
13. Dobbs, R. A. and Cohen, J. M. "Carbon Adsorption Isotherms for Toxic Organics," Municipal Environmental Research Laboratory, Office of Research and Development, U.S. Environmental Protection Agency, EPA-600/8-80-023 (April 1980).
14. Kamlet, M. J., Doherty, R. M., Abboud, J. M., Abraham, M. H., and Taft, F. W. "Linear Solvation Energy Relationships: XXXVI. Molecular Properties Governing Solubilities of Organic Nonelectrolytes in Water," *J. Pharm. Sci.* 75 (1986).
15. Leo, A. J. and Weininger, D. "CLOGP3," Version 3.2, Medicinal Chemistry Project, Pomona College, Claremont, CA (1984).
16. Nirmalakhandan, N. and Speece, R. E. *Environ. Sci. Technol.* 22:606 (1988).
17. Nirmalakhandan, N. and Speece, R. E. "Prediction of Aqueous Solubility of Organic Chemicals Based on Molecular Structure. II. Applications to PNAs, PCBs, PCDDs, etc.," *Environ. Sci. Technol.* 23:708 (1989).
18. Syracuse Research Corporation: *Environmental Fate Data Bases* (1987).

CHAPTER 5

Competitive Adsorption of Several Organics and Heavy Metals on Activated Carbon in Water

Z. P. Jiang, Z. H. Yang, J. X. Yang, W. P. Zhu, and Z. S. Wang

INTRODUCTION

Adsorption with activated carbon has been one of the most useful techniques in water treatment. While in the past, activated carbon was predominantly used to remove odor- and color-producing molecules in water,[1] recent experimental results have been reported for the removal of inorganic chemicals with it.[2] Since multipollutant systems encountered in water treatment contain both inorganics and organics and inorganics can affect the adsorption of organics onto activated carbon, it is important to study their competitive adsorption. In this chapter, the competitive adsorption of organic pollutants (phenol, humic acid) and inorganic pollutants (Pb, Cd, and Cr) are studied.

EXPERIMENTAL PROCEDURES

Materials

Adsorbent

ZJ-15 granular activated carbon (GAC) manufactured in China was used. It was pretreated as follows: GAC → rinsing with ultrapure water → air-dried → ground (<320 mesh) → preserved in desiccator.

Adsorbates

Phenol, $Pb(NO_3)_2$, $Cd(NO_3)_2$, and $K_2Cr_2O_7$ (analytical reagent grade) were used to prepare solutions of 1000 mg/L phenol, Pb(II), Cd(II), and Cr(VI), respectively.

Humic acid (biochemical reagent grade, 10 g) was dissolved in a hot solution of 30% NaOH. The solution was cooled and filtered. Then concentrated sulfuric acid was added to adjust the pH to 7.0, and the solution was diluted to 1000 mL.

The total organic carbon (TOC) was determined as milligrams of TOC per liter. The adsorbate solutions were diluted with ultrapure water to the range of concentration needed, and phosphate buffer was added for adjusting pH to the required value before each experiment.

Analytical Methods

The concentrations of phenol and humic acid were measured with a UV spectrophotometer at 270 and 260 nm, respectively. All metal concentrations were measured by atomic absorption spectrophotometry.

Experimental Method

The adsorption isotherms were developed using the completely mixed-batch reactor bottle-point technique. Accurately weighed amounts of activated carbon were added to a group of 250-mL stoppered Erlenmeyer flasks. Various carbon dosages were designed to cover the range of equilibrium solute concentration of interest. One bottle was left to serve as a blank. Aliquots (100 mL) of the prepared solutions (single solute or bisolute) were placed into each flask. Then, each stoppered flask was placed on a shaker for a given time to reach equilibrium. The content of each flask was then vacuum filtered through 0.45-μm Millipore filters for solid-liquid separation, and the residual adsorbate concentration in the filtered solutions was determined.

After equilibrium, the relationship between adsorption capacity and equilibrium concentration is

$$q = \frac{(C_o - C_e)V}{M}$$

where q = adsorption capacity (mg/g)

C_o = initial concentration (mg/L)

C_e = equilibrium concentration (mg/L)

M = carbon dosage (g)

V = solution volume (L)

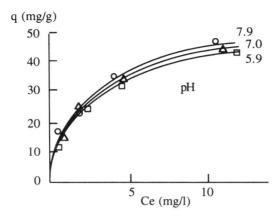

FIGURE 1. Phenol (single solute system, at different pHs).

The equilibrium capacities of the pollutants in single solute and bisolute systems at pH 5.9, 7.0, and 7.9 were investigated in this paper. The pH of water or wastewater to be treated is usually near this neutral range. All experiments in this study were carried out at room temperature (18 to 20°C).

In this chapter, all mechanisms for the removal of chemicals from the system will be considered as the value used for the Y-axis (milligrams per gram). The difference of concentrations of, for example cadmium, before an experiment will have the residual left (Ce) subtracted from it, even if cadmium precipitates from the solution as $Cd(OH)_2$.

RESULTS AND DISCUSSION

Single Solute Systems

Phenol

Figure 1 shows the adsorption isotherms of phenol at different pHs in the single solute system. The figure shows that the adsorption capacity of phenol increases at higher pH, but the change is small. Phenol (pK_a = 9.9) does not ionize to its ionic form in the experiment as pH < pK. The number of adsorption sites available for the phenol molecular form does not vary greatly,[3] so the adsorption capacity of phenol does not change very much.

Humic Acid (HA)

Figure 2 is the adsorption isotherms of HA at the varied pHs in the single solute system. Adsorption capacity of HA decreases as pH increases. This is

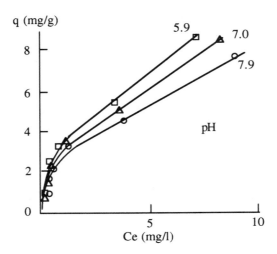

FIGURE 2. Humic acid (single solute system, at different pHs).

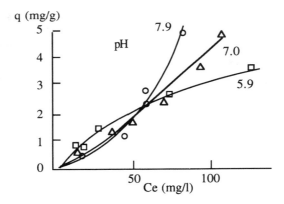

FIGURE 3. Pb (single solute system, at different pHs).

attributed to the complicated structure and property of HA. Meanwhile, the affinity between HA and activated carbon is associated with the pH value. At lower pH, the structure of HA takes the shape of a filament or a collection of filaments. At higher pH, the structure becomes like a net or ball, similar in structure to a sponge. Thus at high pH, humic acid demonstrates typical properties of a polyelectrolyte. Obviously, uncharged HA at lower pH is easier to adsorb. Moreover, the solubility of HA decreases at lower pH, causing possible precipitation on activated carbon. As a result, it seems that the adsorption capacity of HA increases as pH decreases.

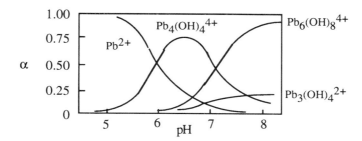

FIGURE 4. Fraction of the species (α) of Pb(II) at different pHs.

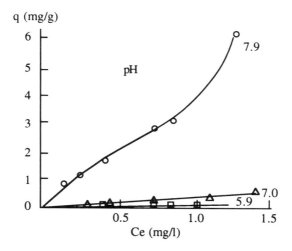

FIGURE 5. Cd (single solute system, at different pHs).

Pb(II)

The adsorption isotherms in Figure 3 show that the adsorption capacity of Pb(II) increases dramatically with increasing pH. The increasing pH brings about not only the precipitation of $Pb(OH)_2$, but also the production of some polynuclear cationic hydroxycomplexes[4,5] such as $[Pb_6(OH)_8]^{4+}$, $[Pb_4(OH)_4]^{4+}$, and $[Pb_3(OH)_4]^{2+}$ as well as $Pb(OH)^+$ and $Pb(OH)_2$. Because of the precipitate and complexes, the concentration of free ionic Pb^{2+} in total Pb(II) decreases gradually (Figure 4). So, the equilibrium concentration of Pb(II) in the filtrate decreases. The adsorption of lead onto activated carbon increases.

Cd(II)

The adsorption capacity of Cd(II) is greatly affected by pH. Figure 5 shows that at a pH lower than 7.0, Cd(II) is adsorbed by activated carbon slightly,

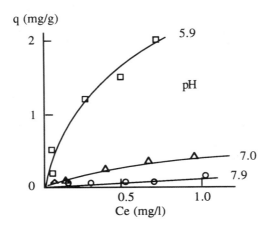

FIGURE 6. Cr (single solute system, at different pHs).

while at pH 7.9, the adsorption capacity increases markedly. Generally, the effect of activated carbon on removing free metal ions is poor. However, if the state of metal is changed, such as in complexation or precipitation, the heavy metal adsorption might be improved. The adsorption of Cd(II) is just an illustration of that point. At pH <7, free ionic Cd^{2+} is the predominant species of Cd(II), so the adsorption onto activated carbon is small. As the pH increases, $Cd(OH)_2$ will precipitate and that will greatly improve the adsorption.

Cr(VI)

Figure 6 shows that the adsorption capacity of Cr at pH 5.9 is the greatest for the range of pHs studied. Some Cr(VI) is reduced to Cr(III) at that pH. $Cr_2O_7^{2-}$ and $Cr(OH)^{2+}$ can be adsorbed onto the activated carbon at two types of charge sites. When pH >6, the reduction is lowered. Additionally, $Cr_2O_7^{2-}$ will change into CrO_4^{2-}, the latter not being better than the former with respect to the adsorption. Therefore, the adsorption capacity of Cr decreases remarkably. This demonstrates the adsorption and removal of Cr(VI) is improved under acidic conditions.

Bisolute Systems

Phenol–Pb(II)

Figures 7 and 8 are the adsorption isotherms of phenol and Pb(II), respectively, in the bisolute system.

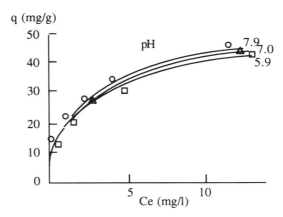

FIGURE 7. Phenol (phenol-Pb system, at different pHs).

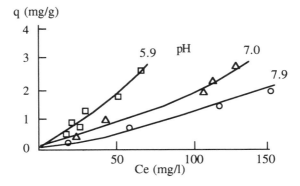

FIGURE 8. Pb (phenol-Pb system, at different pHs).

Comparing Figure 7 with Figure 1, it shows that under the same condition the adsorption capacity of phenol in the bisolute system is almost the same as in the single solute system. Adsorption of phenol is unaffected by Pb species. Because activated carbon is a nonpolar adsorbent onto which physical adsorption occurs prior to chemical adsorption, the organic (phenol) that is mainly physically adsorbed prior to the inorganic (Pb) that is mainly chemically adsorbed occupies the nonpolar adsorption sites. Thus, the adsorption capacity of phenol varies only slightly.

On the other hand, comparing Figure 8 with Figure 3, it shows that the variation of the adsorption capacity of Pb at various pH in the bisolute system is quite different from the single solute system. At pH 5.9, adsorption of Pb in the bisolute system is improved slightly due to the existing phenol. Because the free

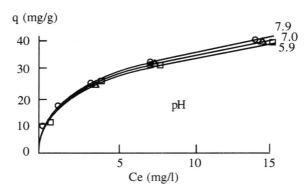

FIGURE 9. Phenol (phenol-Cd system, at different pHs).

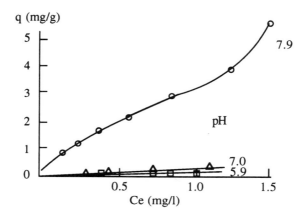

FIGURE 10. Cd (phenol-Cd system, at different pHs).

ion Pb^{2+} is a greater fraction of all species of Pb(II) at that pH (Figure 4), it can combine with the phenol adsorbed onto the activated carbon to form $Pb(OC_6H_5)_2$. Consequently the adsorption capacity of Pb increases. While under neutral or weak basic conditions (pH 7.0 and 7.9), the major species of Pb(II) are $Pb_4(OH)_4^{4+}$ and $Pb_6(OH)_8^{4+}$. They are difficult to combine with the phenol adsorbed on activated carbon. The adsorption of Pb is resisted by phenol which acts as a competitive compound so that the adsorption capacity of Pb decreases remarkably. In addition, precipitation and hydroxy-complexation still occur in the bisolute system.

Phenol–Cd(II)

The adsorption isotherms of phenol and Cd(II) in the bisolute system are shown in Figures 9 and 10, respectively. Adsorption capacities of both phenol

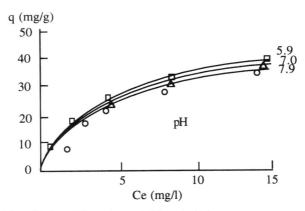

FIGURE 11. Phenol (phenol-Cr system, at different pHs).

and Cd(II) at various pH are lower than in the single solute system. This illustrates competition for the same sites on activated carbon. Comparing Figure 10 with Figure 5, it also shows that at pH 7.9 the adsorption capacity of Cd(II) in the bisolute system decreases much more than in the single solute system. This can be explained as follows. In the single solute system, the adsorption capacity of Cd(II) increases with increasing pH due to the formation of $Cd(OH)_2$ precipitation. However, in the bisolute system, some Cd^{2+} combines with phenol, which is easily soluble, to form $Cd(OC_6H_5)_2$. The formation of $Cd(OH)_2$ precipitation decreases, and as a result, the adsorption of Cd(II) on activated carbon is inhibited.

Phenol–Cr(VI)

At different pHs, the adsorption of both phenol and Cr in the bisolute system decreases slightly compared with that in the single solute system, but the shape and tendency of the isotherms do not change remarkably (Figures 11 and 12). Phenol cannot combine with either the anion $Cr_2O_7^{2-}$ or CrO_4^{2-}. There is only simple competitive adsorption between phenol and Cr(VI) in the system.

Humic Acid–Cd(II)

Figure 13 presents adsorption isotherms of HA at various pHs in the bisolute system. It shows that the adsorption does not change very much compared with Figure 2. However, it is notable that all of these curves do not pass through the coordinate origin and there are intercepts at the abscissa. This phenomenon cannot be explained by the classic adsorption isotherm models, such as Langmuir and Freundlich. In the presence of Cd(II), adsorption of humic acid is changed

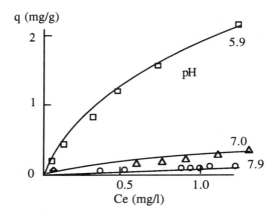

FIGURE 12. Cr (single solute system, at different pHs).

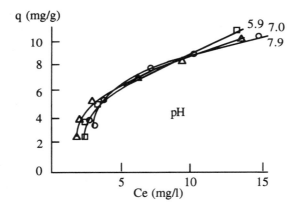

FIGURE 13. HA (HA-Cd system, at different pHs).

perhaps due to variation of structural state and molecular weight of the humic acid. Infrared spectrum analysis (see Figure 14)[6] has shown that the presence of Ca^{2+} results in the occurrence of new peaks on the HA spectrum. Comparison of Figure 15 with Figure 5 shows that humic acid can remarkably inhibit the adsorption of Cd(II).

Humic Acid–Cr(VI)

Figures 16 and 17 show adsorption isotherms of the bisolute system at different pHs, respectively, which are similar to the HA–Cd(II) system. In both cases,

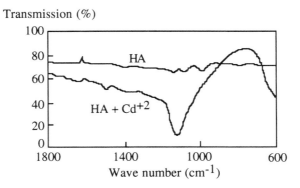

FIGURE 14. The infrared adsorption spectrum of HA.

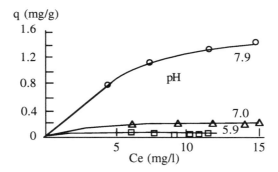

FIGURE 15. Cd (HA-Cd system, at different pHs).

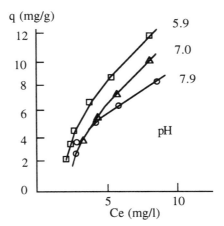

FIGURE 16. HA (HA-Cr system, at different pHs).

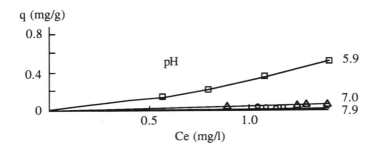

FIGURE 17. Cr (HA-Cr system, at different pHs).

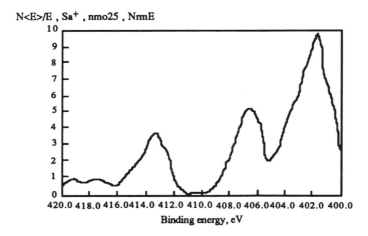

FIGURE 18. Cd in the pure sample.

adsorption capacity of HA does not decrease compared with the single solute system. But these isotherms also do not go through the origin and have the intercepts at the abscissa. In the presence of HA, the adsorption capacity of Cr(VI) remarkably decreases and in the meantime, the curvature of the isotherms is also changed.

All of these indicate that there are more complicated chemical reactions in these bisolute systems which consist of the high polymer, such as humic acid, and heavy metals such as Cd and Cr. These materials complicate simple interpretation of adsorption.

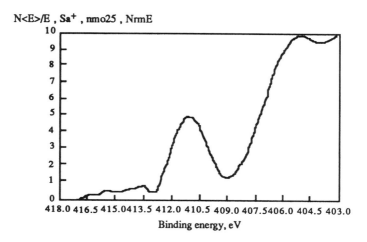

FIGURE 19. Cd in the adsorbed sample (single solute system, at pH 7.9).

X-Ray Photoelectron Spectrum

These experimental results show that in the adsorption process of heavy metals on activated carbon, their chemical form and valences of the metal may be changed. In order to confirm this conclusion, X-ray photoelectron spectrum analysis was conducted by selecting Cd(II) and Cr(VI) as representatives of cation and anion, respectively.

Figure 18 shows the Cd photoelectron spectrum for pure samples of $Cd(NO_3)_2$. The Cd spectrum peak is at 413.4 eV in this figure. Deducting charging effect (2.0 eV), the net value of the Cd spectrum peak is 411.4 eV.

Figures 19 and 20 present photoelectron spectra of Cd species adsorbed on activated carbon at pH 7.9 in single solute and bisolute systems (phenol–Cr), respectively. The peaks are located at 411.2 and 410.2 eV, respectively, which indicates that the binding energy of Cd adsorbed on activated carbon shifts to a lower energy level compared with the pure sample. This is an indication that Cd is to some extent reduced chemically.

Figure 21 is the spectrum of Cr for the $K_2Cr_2O_7$ standard. The Cr peak is located at 582.0 eV. Deducting the charging effect of 2.6 eV, the net peak value is 579.4 eV. The photoelectron spectra of adsorbed Cr on activated carbon at pH 5.9 in the single solute and bisolute systems are shown in Figures 22 and 23, respectively. Peaks are located at 578.2 and 577.8 eV, respectively. In these cases, the binding energy of Cr adsorbed on activated carbon tends to decrease the energy level compared with the pure sample, demonstrating a partial reduction of Cr(VI) to Cr(III). Additionally in Figures 22 and 23, peak overlap occurs and become broader, indicating that there are multiple chemical species of Cr on the activated carbon. Various functional groups on activated carbon may take important roles.

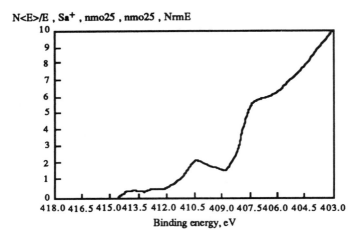

FIGURE 20. Cd in the adsorbed sample (phenol-Cd system, at pH 7.9).

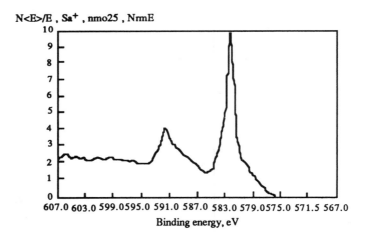

FIGURE 21. Cr in the pure sample.

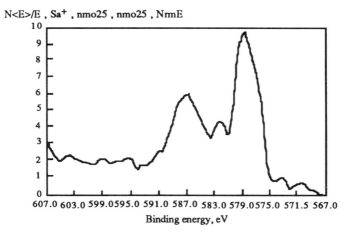

FIGURE 22. Cd in the adsorbed sample (single solute system, at pH 5.9).

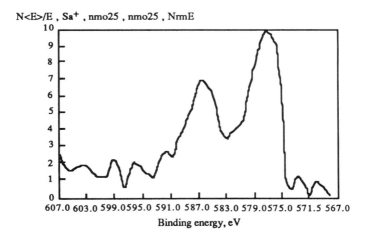

FIGURE 23. Cr in the adsorbed sample (phenol-Cr system, at pH 5.9).

CONCLUSION

1. During activated carbon adsorption, the effect of pH on phenol adsorption is caused by phenol dissociation. In natural water, because the pH is commonly in the range of 6 to 8, phenol occurs in the neutral form because the pH < pK_a. In this case, pH does not significantly affect the phenol adsorption, while it considerably affects the humic acid adsorption apparently through changing dispersion-aggregation, properties of humic acid. The solution pH also dramatically affects the adsorption of heavy metals, e.g., Pb, Cd, and Cr, because the variation of pH changes the valences and species of these heavy metals.
2. The adsorption behavior of activated carbon is quite different between organics (phenol and humic acid) and inorganic compounds (Pb, Cd, and Cr). For organics it is mainly a physical adsorption process, while for heavy metals, it is a chemical adsorption. For the bisolute system of organics-metal tested in this paper, both physical and chemical adsorption occur together and the former is more important. Therefore, the adsorption of organics is hardly affected by the heavy metals, while phenol and humic acid significantly change the heavy metal adsorption. This conclusion has also been drawn from the adsorption experiments for trisolute systems such as phenol–Cr–Cd, phenol–Pb–Hg and humic acid–Cr–Cd, etc.[7]
3. The adsorption of heavy metals (Pb, Cd, and Cr) on activated carbon may be affected by ion exchange, oxidation and reduction, complexation, precipitation and competition, etc., simultaneously. Due to these effects, the chemical form and valence of heavy metal will change in the adsorption system to create a complex adsorption isotherm. Among these factors, the adsorption characteristics of the different complexes play an important role.
4. X-ray photoelectron spectroscopy has confirmed that multiple species are involved in the adsorption of heavy metal on activated carbon.

REFERENCES

1. Suffet, I. H. and McGuire, M. J. *"Activated Carbon Adsorption of Organics from Aqueous Phase,"* Vol. 1 (Ann Arbor, MI: Ann Arbor Science, 1980).
2. Anderson, M. A. and Robin, A. J. "Adsorption of Inorganic at Solid-Liquid Interfaces" (1985).
3. Jiang, Z. P. "The Competitive Adsorption of Various Phenols in Water onto Granular Activated Carbon," *J. Water Treat.* 2(2) (1987).

4. Snoeyink, V. and Jenkins, D. *Water Chemistry,* (New York: John Wiley & Sons, 1980), p. 214.
5. Tan, T. C. and Teo, W. K. *J. Water Res.* 21(10):1183 (1987).
6. Jiang, Z. P. and Liao, M. J. "A Study on Activated Carbon Adsorption Equilibrium of Humic Acid," *J. Water Treat.* 14(5) (1988).
7. Yang, J. X. MS Thesis (in Chinese), Tsinghua University, China (1989).

CHAPTER 6

Coagulation-Flocculation of Minerals Using Al,Fe(III) Salts. What Kind of Flocs for What Separation Process?

J. Y. Bottero

ABSTRACT

The aggregation of silica colloids by various aluminum hydroxide polycations (PAC, PACS) and oxy-hydroxide Fe(III)(PIC) has been studied. Both physico-chemical variables like zeta potential, settling velocity, and floc structure on a semilocal scale (10 to 1200Å), by means of small-angle X-ray scattering (SAXS), have been studied as a function of chemical nature, concentration and size of the flocculant, and pH of the silica suspension. Between pH 4.5 to 5.0 and 7.0 to 7.5 the flocs have generally a fractal geometry with a fractal dimension varying from 1.7 to 2.2. Their semilocal effective density, calculated from the distance distribution function obtained from SAXS, was highest at low pH and with PAC as flocculants. With PIC coagulants the fractal structure and effective density were constant and were not modified as pH was varied.

With PAC or PACS, when the pH increased, the density decreased, but remained the highest for PAC flocculants. The settling velocity and the effective density were conversely varied. The mechanisms of the flocculation depended on the reaction pH. At low pH, the charge-neutralization was the main driving force for the flocculation. At neutral pH, both charge-neutralization and bridging were responsible for the formation of less dense flocs.

With PIC, the mechanism was the same: neutralization and bridging. These structures detected in a semilocal range can be dependant on the observation scale: (1) the flocs can be multifractal, the fractal dimension decreases with increasing the scale sizing; or (2) the flocs are "mono" fractal, the fractal dimension is constant whatever the scale sizing. The floc break up could be more effective in the first case, the weak bounds within a floc would be the regions where the structure is changed. In the case of a "mono" fractal floc, the structure is nowhere changed and the break up is not so effective.

INTRODUCTION

Coagulation and flocculation processes are important from industrial and environmental points of view. Characterization and control of the flocs or sludges produced requires control of the properties of the suspensions,[1] floc size distribution,[2] and the development of on-line monitoring equipment.[3-5] The efficiency of the liquid/solid separation by coagulating depends essentially on the chemical and physicochemical and hydrodynamic conditions during mixing and transport of the flocs. These factors determine the structure of the aggregates, i.e., the density and strength the flocs acquire. The first step of the flocculation is the adsorption of the flocculant at the liquid/solid interface. In this step, ionization of the flocculant molecules is of primary importance. Generally nonionic macromolecular flocculants have a weak affinity for solid surfaces and the aggregation mechanism is essentially of a bridging nature.[6,7] The consequence is that the structure of the flocs is loose and fragile. Conversely, the adsorption of cationic organic polyelectrolytes onto negatively charged solid surfaces is strong and the compact structure of the adsorbed layer[8] leads to dense flocs. The mechanism is then the charge-neutralization.[9,10] The size of the flocs can change under the action of fluid stresses. From laboratory experiments with negative latex flocculated by NaCl, it was found that the size of flocs was proportional to shear rate γ raised to a power which is a function of the fractal dimension.[11] In many industrial processes, aluminum or iron hydroxides are used as flocculants. It has been found that the breaking up of flocs obtained by precipitation of aluminum hydroxide in organic medium is sensitive to the shape of the mixing reactor.[12] Although there are now many theories on kinetics and thermodynamical aspects of coagulation-flocculation,[12,13] but few direct data exist on the structure of flocs obtained by heterocoagulation of negative mineral particles by positive mineral "particles". The coagulation-flocculation of negative colloids by aluminum hydroxyde particles is of outstanding importance in soils and for water purification.

The aim of this paper is to describe the structure of flocs, at the semilocal scale, obtained by mixing silica sols with PAC, PACS, and PIC flocculants. To this end we use small-angle X-ray scattering and try to relate the data to the macroscopic behavior during flocculation and settling.

MATERIALS

Ludox HS 40 (Dupont de Nemours) silica colloids have been used throughout. The mean size is ~14 nm in diameter as deduced from the manufacturer specification and from data obtained by small-angle X-ray scattering or small-angle neutron scattering techniques.[13,14]

The sign of the surface charge determined by measuring the electrokinetic potential was negative over the whole range pH ≥ 2.[15,16] All the experiments were carried out at pH ~4.5 or ~7.5 so that the silica surface charge was always negative. The concentration of the silica sols was 0.8%. Two kinds of aluminum hydroxide flocculants were used.

PAC solutions were obtained by hydrolysis of aluminum chloride with sodium hydroxide following a procedure described previously.[17] The initial hydrolysis ratio of the aluminum hydroxides was R = (OH)/(Al) = 2, 2.5. The initial aluminum concentration was 0.1 M and the pH was varied from ~4.0 (R = 2) to 5.0 (R = 2.5). So the aluminum species[17,18] present are mainly monomers and the polycation $Al_{13}O_4OH_{24}(H_2O)_{12}^{7+}$. The Al_{13} ions build fractal aggregates in R = 2.5 solution.[19] For R = 2.5 the fresh aggregates contained ~69 Al_{13} and have a fractal dimension D_f = 1.43 which corresponds to a very linear "polymer" (Figure 1).

The second aluminum flocculant used was an industrial polyaluminum chlorosulfate or PACS (ATOCHEM) with 3% of SO_4^{2-} ions. The pH of the mother solution was ~2.2. The species present in the mother solution were essentially monomers and chains of octahedral aluminum.[22]

The Fe(III) flocculants are obtained by hydrolyzing $FeCl_3 6H_2O$ with NaOH until ratios of R = 1.0, 2.0, and 2.5 are obtained. The initial concentration of the mother solutions was 0.1 M. The Fe(III) speciation in these solutions is essentially monomers, dimers, and small clusters of 10 to 15 Fe atoms, the local structure of which was detected by X-ray absorption spectroscopy.[23] This cluster is built with octahedral Fe linked by edges and corners. It is an oxy-hydroxy ferric cluster. The Fe flocculants are named poly iron chloride or PIC.

EXPERIMENTS

Flocculation Experiments

Experiments were carried out by rapid-mixing of a silica sol with a given amount of flocculants at a constant pH of 4.5 or 7.5 following a jar-test procedure which consists of stirring the suspension in a 1-L vessel at 250 rpm during 2 min and 40 rpm during 20 min; the experimental chemical conditions are given in Table 1. The pH was adjusted using concentrated NaOH. The flocs were allowed

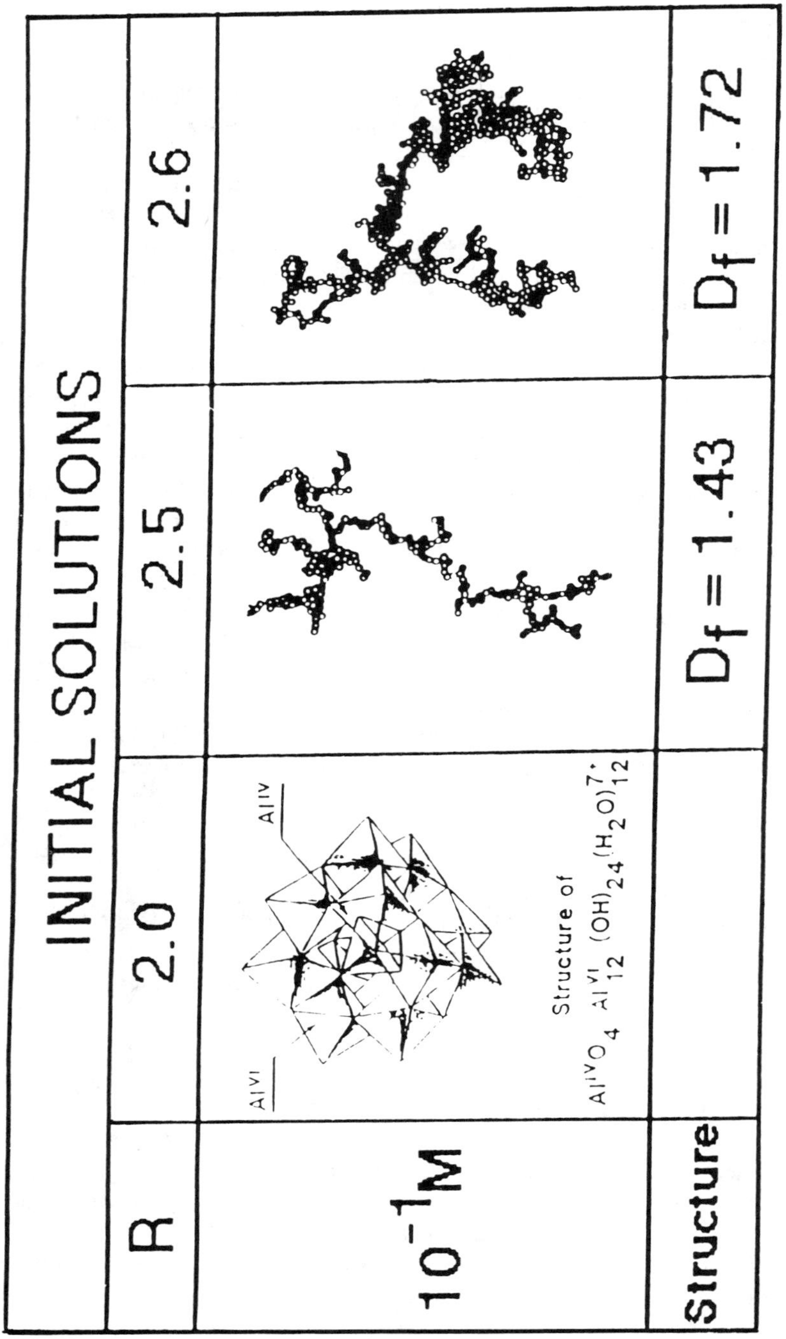

INITIAL SOLUTIONS			
R	2.0	2.5	2.6
10^{-1} M	Structure of $Al^{IV}O_4 \; Al^{VI}_{12}(OH)_{24}(H_2O)^{7+}_{12}$		
Structure		$D_f = 1.43$	$D_f = 1.72$

FIGURE 1. Structures of PAC flocculants used.

Table 1. Dosages of Flocculants after Mixing with 0.8% Silica Suspension

	Flocculant Dosage (mol/L)					
	10^{-3}	2×10^{-3}	4×10^{-3}	6×10^{-3}	8×10^{-3}	10^{-2}
PAC						
R = 2	+	+	+	+	+	+
R = 2.5	+	+	+	+	+	+
PACS	+	+	+	+	+	+
PIC						
R = 1.0			+	+		
R = 2.0			+	+		
R = 2.5			+	+		

to settle in a 50-cm³ glass tube, vertically fixed in a carefully thermostated area and kept out of vibrations. The height of the sediment front was measured vs time.

The electrokinetic potential of the flocs vs flocculant concentration was measured at pH ~4.5 and ~7.5 using a Laser Zee meter electrophoresis apparatus (Pen Kem Inc). The flocculation efficiency vs flocculant dosage was followed by measuring (1) the turbidity of the suspensions after 20 min of slow stirring and before settling, (2) the turbidity of the supernatant after settling the suspension during 1 hr, and (3) the residual silica content (PIC). The optimum flocculation concentration (ofc) was determined as the lowest flocculant dosage (moles per liter) corresponding to the lowest turbidity.

During the dilution of hydrolyzed Al or Fe in silica suspension at pH ~7.0 to 7.5, further hydrolysis of previous species can proceed.[27] Al NMR experiments were carried out on Al flocculants precipitated at pH 7.5 in order to detect possible structural change at the local scale during hydrolysis. The spectra were obtained at 132 MHz at magic angle spinning (~54°C and 3.8 KHz). The sequences, recording method, and the method for calculating the Al_{13} content of the precipitates were described earlier and will not be repeated.[19]

Small-Angle X-Ray Scattering

Experiments were carried out on flocs obtained on both sides of the ofc and at the ofc for each one of the systems investigated in Table 1. The scattering from the suspensions is due to the silica colloids. The scattering by Al or Fe species ($<10^{-2}$ M) is the same as the scattering by the solvent.

X-Ray Data Recording

Synchrotron radiation of the DCI storage ring of LURE (Université de Paris-sud, Orsay) was used for its high intensity associated with point collimation.

Very small angles were reached and the collected data covered the Q-range from $2.8 \cdot 10^{-3}$ Å^{-1} to $3.6 \cdot 10^{-1}$ Å^{-1} (Q is the scattering vector amplitude; $Q = 4\pi \sin\theta/\lambda$ with 2θ as the scattering angle and λ the wavelength 1.6 Å).

X-Ray Data Analysis

The scattered intensity by aggregates I(Q) is

$$I(Q) = KI_0(Q)G(Q) \tag{1}$$

where K is a scale constant of the experiment
 $I_0(Q)$ is the scattering intensity by a subunit silica particle
 G(Q) is the interference function depending on the silica interparticle arrangement within an aggregate

The interference function G(Q) is only valid within the Q-range:

$$1/L \leq Q \leq 1/R_0$$

where L is the characteristic length of the aggregates
 R_0 is the characteristic size of the subunit

If the aggregates are mass fractals:[13,21,24,25]

$$G(Q) \sim Q^{-D_f} \tag{2}$$

where D_f is the fractal dimension which can be calculated from log I(Q) vs log(Q) plot within the convenient Q range.

The scattered intensity per aggregate is given by:

$$I_N(Q) = I(Q)/PO \tag{3}$$

where PO is the scattering power of the sample which corresponds to the content of the scattering:

$$PO = \int Q^2 I(Q) ds \tag{4}$$

The correlation function for an aggregate in the direct space is $\gamma(r)$ which is calculated from the Fourier transform of $I_N(Q)$:

$$\gamma(r) = \pi^2/2 \int Q^2 I_N(Q)(\sin Qr/Qr) dQ \qquad (5)$$

and the distance distribution function $P(r)$ is the second moment of $\gamma(r)$:

$$P(r) = r^2 \gamma(r) \qquad (5')$$

The average volume of matter Vm within an aggregate is computed from the relation:

$$4\pi \int P(r) dr = Vm \qquad (6)$$

Experimentally the aggregates are often larger than the accessible scale. The present SAXS experiments have a limited Q range: $2.8 \ 10^{-3} \ \text{Å}^{-1} \leq Q \leq 3.6 \ \text{Å}^{-1}$. In other words, due to the limitation of Q_{min}, the aggregates are "seen" through a window with a diameter $D_w = \pi/Q_{min} \sim 1150$ Å. The scattering curves at large Q are related to the subunit particle; this region must be extrapolated up to $Q \sim 8 \ \text{Å}^{-1}$ in order to calculate the Fourier transform.

The average size of the subunit is calculated from the scattering of a diluted silica Ludox suspension.[14] Under such conditions where D_w is lower than L, $P(r)$ leads to Vm values which correspond to the experimental average volume of matter delimited within the window diameter D_w. The Vm/V_0 ratio, where V_0 is the volume of the subunit silica particle, gives the statistical aggregated particle average number N per experimental unit volume. Calculating D_w from each experimental condition, it is possible to compare the average effective floc densities obtained in various physical and chemical conditions. The density ρ is given by:

$$\rho = Nm_0/V_w \qquad (7)$$

where $V_w = \pi D_w^3/6$ is the volume corresponding to the window diameter D_w, N the silica particle average number in the V_w experimental volume, and Nm_0 the mass of silica in V_w. The mass of one silica particle m_0 is calculated from:

$$M_0 = \rho_{silica} \cdot V_0 \qquad (8)$$

From previous studies $V_0 = 2.5 \ 10^{-18} \ \text{cm}^3$ as calculated from SAXS technique.[14]

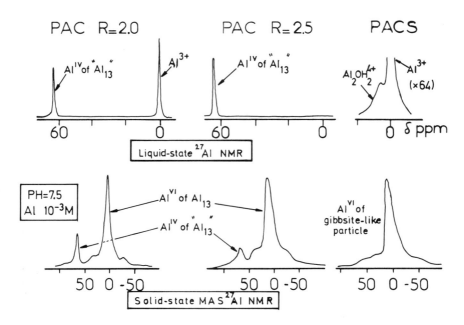

FIGURE 2. Liquid-state ^{27}Al NMR of PAC and PACS initial solutions and solid-state ^{27}Al NMR of flocs precipitated from PAC (R = 2.0 and 2.5), PACS at pH 7.5 and (Al) $\sim 10^{-3}$ M.

RESULTS

Chemistry of the Flocculants at Neutral pH

The *solid and liquid state ^{27}Al NMR* patterns indicate that the local-range order is not significantly modified by the dilution and pH increase. The aggregates formed at pH 7.5 from R = 2, 2.5 are largely constituted of Al_{13} polycations as proved by the peak of resonance at 62.5 ppm from the reference $(Al(H_2O)_6^{3+})$ (Figure 2) characteristic of the tetrahedral aluminum at the center of the Keggin Al_{13} structure.[19] PACS never contain these kind of polymers and the aggregates are only constituted of octahedral aluminum probably linked by SO_4^{2-} ions. The Al_{13} content in the aluminum precipitates can be calculated from NMR patterns.[19] The calculation shows that the Al_{13} concentration is high whatever the initial R values and age of the precipitates at neutral pH values (Table 2). This means that the local structure of aluminum precipitates obtained from initial R = 2, 2.5 values is essentially that of Al_{13} polycation. The electrostatic charge of aluminum flocculants changes with pH because the surface of such

Table 2. Evolution of the Al_{13} Content in Precipitates from R = 2.0 and 2.5 at Neutral pH vs the Aging Time

pH	6.5		7.5	
Age of precipitates	10 min	1 hr	10 min	1 hr
R = 2.0	~100%	~80%	~100%	~80%
R = 2.5	~100%	~80%	~80%	~60%

Table 3. ofc for the Different Flocculants and Various pHs

	pH	
	4.5	7.5
PAC		
R = 2.0	4×10^{-3}	8×10^{-3}
R = 2.5	4×10^{-3}	8×10^{-3}
PACS	2×10^{-3}	8×10^{-3}
PIC	4×10^{-3}	6×10^{-3}

particles has the same behavior as oxides. The surface charge evolution of isolated and aggregated Al_{13} has been measured in the pH range 4 to 8.4 (ZPC).[22,26] At pH ~4.5 the charge of Al_{13} is maximum and corresponds to 1.5 to 1.6 H^+/nm^2, at pH ~7.5 the charge is 0.5 proton/nm^2. The surface charge of WAC has not been adequately studied, but the ZPC is ~8.2 as deduced from electrophoretic mobility experiments in deionized water and NaCl electrolyte 10^{-3} M indicating a net positive surface charge at the different experimental pH conditions used.

Flocculation and Electrophoresis Results

The ofc depends essentially on the pH of the flocculation (Table 3). The sedimentation data at ofc for PAC and PACS at pH 4.5 and 7.5 show that the settling velocity is the highest with PACS followed by PAC (R = 2) and PAC (R = 2.5) flocculants. Increasing pH decreases the settling velocity (Table 4). The velocity (V) can be related with the floc sedimentation coefficient S through:

$$S = V/A \qquad (9)$$

where A is the acceleration of the particles

In the case where the sample is at rest: A = g (the acceleration of gravity). For fractal aggregates,[28-32] S can be related to the average hydrodynamic diameter

Table 4. Settling Velocity of Flocs at ofc vs the Nature of the Flocculant at pH = 4.5 and 7.5

	R = 2.0		R = 2.5		WAC		PIC	
pH	4.5	7.5	4.5	7.5	4.5	7.5	4.5	7.5
V (cm/sec)	4.05×10^{-4}	5.7×10^{-5}	1.62×10^{-4}	7.5×10^{-5}	1.03×10^{-3}	4.05×10^{-4}	ND	ND

d_H (which is not far from the cluster diameter for fractal dimension D_f between 1.7 and 2.2[32]) through Stoke's law:

$$V = K \cdot g \cdot \rho \cdot d_H^2 / \eta \tag{10}$$

where η is the dynamic viscosity of the medium (water)
    K is a coefficient which takes into account the friction coefficient of the flocs and the Reynolds number

Using data obtained in previous experiments carried out with alum and Kaolinite[27] or PAC and Montmorillonite,[33] the product Kg/η is assumed to be 2880. Then Equation 10 is

$$V \sim 2880 \, \rho \, d_H^2$$

Therefore, it is possible to evaluate the evolution of d_H from the settling velocity experiments (Table 4) and the calculation of ρ from SAXS using Equations 6 and 7.

The zeta potential evolution vs initial flocculant dosage is displayed in Figure 3. At pH 4.5 the iso-electric-point (i.e.p.) is attained for ofc values with Al coagulants. This behavior has been previously observed in the case of Montmorillonite/PAC systems.[42] At pH 7.5 the flocs always have a negative surface charge indicating a lower interaction between PAC, PACS, or PIC and the silica surface.

Small-Angle X-Ray Scattering

Two kinds of curves are plotted. The Log $I_N(Q)$ vs Log(Q) (Figure 4) and distance distribution function P(r) (Figure 5). All the curves exhibit the same shape at large Q, due to the scattering of silica subunits (Figure 4). At the smallest Q range, the scattering depends on the aggregation. In these cases, following Equation 2, the slope of the curves, calculated using the least-square fitting method, is $-D_f$ which is the fractal dimension of flocs. Such a shape is common in the different systems studied.

The distance distribution functions P(r) show large differences following the nature of the flocculant. Figure 5 displays P(r) functions obtained at ofc at pH 4.5 and 7.5. The intensity of P(r), whatever the pH, decreases respectively in the case of flocculation with PAC (R = 2.5, R = 2.0), PACS, and PIC where it is constant. The volume of material V_m (Equation 6), i.e., the number of silica particles (N) per unit volume experimentally accessible, and ρ (Equation 7) also are calculated. The data are given in Table 5. Large variations are exhibited not only for N and ρ, but also for D_f.

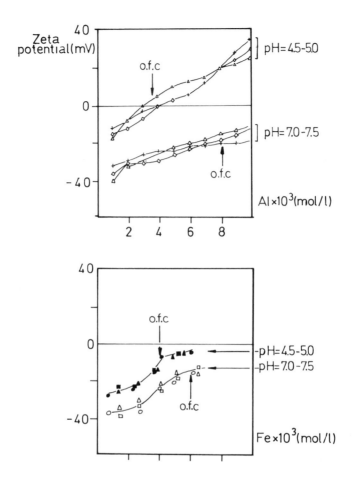

FIGURE 3. Evolution of zeta potential vs flocculant concentration at different flocculation pH: ($\triangle$) = PACS; (+) = PAC R = 2.0; ($\diamond$) = PAC R = 2.5; ($\triangle$,▲) = PIC R = 1.0; ($\circ$,●) = PIC R = 2.0; ($\square$,■) = PIC R = 2.5. [Adapted from Bottero, J. Y. et al., *Langmuir*, 6:596 (1990).]

DISCUSSION

The nmr data show that the subunit of the flocculants PAC are, whatever the pH, formed by Al_{13} polycations. The differences between the flocculants used are the size, the charge (cationicity), and the conformation (small isolated clusters or "linear" aggregates) (Figure 1). The charge decreases and the size increases as the pH increases. The zeta potential of flocs vs Al and Fe concentration at pH = 4.5 and 7.5 (Figure 3), combined with the determination of ofc, by

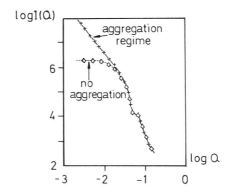

FIGURE 4. logI(Q) vs logQ plots with and without aggregation.

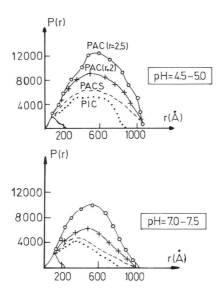

FIGURE 5. Distance distribution function P(r) vs r(Å) of aggregated silica particles within the flocs function of flocculants and pH: (○) PAC R = 2.5, (+) PAC R = 2.0, (--) PACS, (···) Pic R = 1.0, 2.0, or 2.5.

analogy with cationic organic macromolecules,[10,35,36] seem to correspond to two different mechanisms of flocculation.

At pH = 4.5, ofc values obtained with PAC or PACS correspond to the i.e.p.; the mechanism is in large part due to charge neutralization as described by other workers.[35,36] Evidently this model has been developed to explain the flocculation

Table 5. Fractal Dimension D_f, Silica Particle N Within V_w, Effective Floc Density ρ (g/cm^3) and Floc Size at ofc vs the Nature of Flocculant and Flocculation pH

	PAC (R = 2.0)		PAC (R = 2.5)		PACS		PIC	
pH	4.5	7.5	4.5	7.5	4.5	7.5	4.5	7.5
D_f	2.06	1.75	2.24	2.06	2.00	1.92	~1.7	~1.7
N	18	12	26	20	15	10	~10	~10
ρ 10^2 (g/cm^3)	7.83	5.21	11.3	8.7	6.52	4.35	2	2
dH (μm)	13.3	6.16	7.0	5.5	23.4	18.0	ND	ND

Adapted from Bottero, J. Y. et al., *Langmuir*, 6:596 (1990).

of particles by very linear ionic organic polymers or by small cations. In the present case, the molecules of flocculants which lead to a similar flocculation (i.e.p. corresponds to ofc) are either small polycations (for example, isolated Al_{13}^{7+}) or not very branched "polymer" with a low fractal dimension $D_f = 1.4$. This can signify that the fractal aggregates would be adsorbed in a "flat" conformation at the silica–liquid interface occupying the maximum number of adsorbing sites due to a high affinity. With the PIC flocculant the i.e.p. is negative at ofc.

At pH = 7.5, the cationicity of the flocculants has decreased because of further surface hydrolysis. The ofc does not correspond to i.e.p; the zeta potential is always negative (Figure 3). The flocculation mechanism can be described according to the classical point of view of interparticle bridging.[35,36] The adsorbed flocculants partly form loops spreading towards the solvent bulk. This behavior indicates that when the pH increases, the kinetics of surface hydrolysis and particle growth of the initial flocculant molecules is faster than the diffusion time to cover colloids such as, for example, to effect minimum stability. It is well known that the rate of hydrolysis of monomeric aluminum is quite high ($\sim 10^{10}$ sec^{-1}) and the rate constants of coagulation are very much lower.[37] But contrary to these last authors who postulated that the bridge formation by means of polymeric aluminum species does not play a significant role, we found that this is only true at low or medium pH where the polymeric species are not extensively hydrolyzed. It does not hold at neutral pH.

The settling velocity data at ofc depends on reaction pH (Table 4). The kinetics are faster at pH = 4.5 than at pH = 7.5 for each flocculant. The settling velocity inceases as the cationicity of the flocculants decreases.

From the values of V (Table 4) and ρ (Table 5), a value of d_H can be estimated. Table 5 shows that d_H values decrease when pH increases for PAC and PACS, increase when PAC and PACS are used whatever the pH. Moreover, d_H increases as the settling velocity increases. This means that V varies essentially as d_H^2. Density and fractal dimension of the flocs increase. Density and size are conversely varied as pH increases. If ofc is exceeded the density of flocs decreases as shown by comparing the P(r) function of flocs obtained at pH 4.5 to 5.0 at

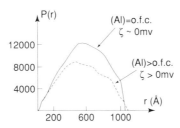

FIGURE 6. P(r) vs r(Å) of flocs obtained at pH = 4.5 to 5.0 for (Al) = ofc (—) and for (Al) > ofc (----).

ofc ($\zeta = 0$ mV; Al = 4 10^{-3} mol/L) and with an excess of flocculant ($\zeta = +20$ mV; Al = 8 10^{-3} mol/L) (Figure 6). The volume of material is the largest at the ofc. The calculation of the particle number N yields: N = 26 at the ofc and N = 20 for Al = 8 10^{-3} mol/L. The fractal dimension D_f calculated from Equation 2 is 2.24 and 2.06, respectively. At pH = 7.5 the same behavior can also be observed.

We have observed an inverse relationship between size and density of the flocs. The density or the fractal dimension increases as the size decreases. This kind of relation has been established in the case of flocculation of clays by alum[27] or Al_{13}.[33] They found by measuring the maximum size L and settling velocity of a large number of flocs and using the Stoke's law, that the effective density increases as size decreases:

$$\rho = AL^{-\alpha}$$

with α varying from ~1.1 to ~1.5 depending on the nature of the flocculants. The fractal geometry of flocs yield: $\alpha = 3 - D_f$ where D_f was varied from 1.9 to 1.5. In the present studied system (silica/aluminum) the parameter α varies according to the aluminum dosage from ~0.70 to ~1.20. More recently, Klimpel and Hogg[39] and Hogg and Ray[40] showed from the flocculation of quartz with organic macromolecules that the floc density–size relationships are complex depending on the size of flocs and the hydrodynamics in the mixing reactor. The fractal geometry is limited in an intermediate floc size range from 50 to 500 μm. The density–floc size plots exhibit a power law which corresponds to fractal dimension values of 1.7 to 2.1. The small flocs are dense and the large ones are very porous.

In our study, the structure of the flocs is characterized on a very limited range size (200 to 120 Å). The calculated effective densities from SAXS data are much higher than those calculated with large flocs.[27,33] That could mean that the dense structures are no longer valid when bigger range size is investigated. The flocs obtained with PAC or PACS could be multifractal and exhibit a decrease of the effective density or of the fractal dimension with size increasing. The consequence of that is that there are at least two mechanisms for the flocculation; first, the formation of dense nuclei where the interactions could not be reversible, and second, the aggregation of the nuclei to form larger flocs through lower energetic interactions. The floc break-up is effective on the second level. The use of various mineral flocculants leads to different structures and certainly different sizes of the nuclei. The break-up should be very different according to the flocculant chemistry.

The variations of D_f and ρ, according to the chemical nature of the flocculants, are in large part due to the high affinity of the Al_{13} species for solids as silicas or silicates[41-44] compared with monomers, other kinds of aluminum colloids, or iron polycations.

CONCLUSION

1. The flocculation of silica colloids by PAC, PACS, and PIC brings about flocs with a fractal geometry at the semi-local range order. One observes a linear relationship between LogI(Q) and LogQ on approximately one decade of distance. The slopes vary and depend on the physical and chemical parameters of the different systems.
2. The fractal dimension and the density of the flocs depend on (a) the chemical nature of the flocculant, i.e., its affinity for the surface of the silica colloids (D_f and ρ are the larger in the case of Al_{13} flocculant); (b) the pH (hydrolysis): D_f and ρ decrease as pH increases; and (c) the flocculant size (ρ drastically decreases, as the size of the flocculant colloids are large compared with that of silica).
3. The mechanism of flocculation depends on the pH. At pH = 4.5, a charge neutralization can be evoked yielding dense flocs. At pH = 7.5, the mechanism is charge neutralization and interparticle bridging, the flocs are less dense.

REFERENCES

1. Tadros, Th. F., *Colloids Surf.*, 18:137 (1986).
2. Cahill, J., Cummins, E. J. W., Staples, E. J., and Thompson, L., *Colloids Surf.*, 18:189 (1986).
3. Gregory, J. and Nelson, D. W., *Colloids Surf.*, 18:175 (1986).
4. Eisenlauer, J. and Horn, D., *Colloids Surf.*, 14:121 (1986).
5. Trimm, H. H., Jennings, B. R., Parslow, K. *Colloids Surf.*, 18:113 (1986).
6. Bottero, J. Y., Bruant, M., Cases, J. M., Canet, D., and Fiessinger, F., *J. Colloid. Interface Sci.*, 124:2, 515 (1988).
7. Griot, O. and Kitchener, J. A., *Trans. Farday. Soc.*, V, 61:1022 (1965).

8. Papenhuijzen, J., Vanderschee, H. A., Fleer, G. J., *J. Colloid. Interface Sci.,* 104:540 (1985).
9. Gregory, J., *J. Colloid. Interface Sci.,* 55:35 (1976).
10. Mabire, F., Audebert, R., and Quivoron, *J. Colloid. Interface Sci.,* 97:120 (1984).
11. Sonntag, R. C. and Russel, W. B., *J. Colloid. Interface Sci.,* 113:399 (1986).
12. Clark, M., Presented at the Annual A.W.W.A. Conference, Denver, 1986.
13. Ramsay, J. D. and Booth, B. O., *J. Chem. Soc.,* Faraday, Trans., 1,79:173 (1983).
14. Axelos, M. A. V., Tchoubar, D., and Bottero, J. Y., *Langmuir,* 5:1186 (1989).
15. Allen, L. H. and Matijevic, E., *J. Colloid. Interface Sci.,* 31:287 (1969).
16. Capelle, P., Ph.D. Université Catholique de Louvain-La-Neuve, 210 p. (1987).
17. Bottero, J. Y., Cases, J. M., Fiessinger, F., and Poirier, J. E., *J. Phys. Chem.,* 84:2933 (1980).
18. Bottero, J. Y., Marchal, J. P., Cases, J. M., Poirier, J. E., and Fiessinger, F., *Bull. Soc. Chim.,* France, I-42:I-43 (1982).
19. Bottero, J. Y., Axelos, M. A. V., Tchoubar, D., Cases, J. M., Fripiat, J. J., and Fiessinger, F. J. *Colloid Interface. Sci.,* 117:47 (1987).
20. Axelos, M. A. V., Tchoubar, D., Bottero, J. Y., and Fiessinger, F. J. *Phys.,* 46:1587 (1985).
21. Axelos, M. A. V., Tchoubar, D., and Jullien, R. *J. Phys.,* 54:1870 (1986).
22. Bottero, J. Y., Tchoubar, D., Cases, J. M., Fripiat, J. J., and Fiessinger, F. *Interfacial Phenomena in Biotechnology and Materials Processing* Y. A. Attia, B. M. Moudgil, and S. Chander, Eds. (New York: Elsevier Sciences, 1988), p. 459.
23. Combes, J. M., Manceau, A., Calas, G., and Bottero, J. Y. *Geochim. Cosmochim. Acta,* 53:583 (1989).
24. Martin, J. E. *J. Appl. Cryst.,* 19:25 (1986).
25. Jullien, R., Botet, R. *Aggregates and Fractal Aggregates* (Singapore: World Scientific, 1987).
26. Rakotowarivo, E., Bottero, J. Y., Thomas, F., Poirier, J. E., and Cases, J. M., *Colloids and Surfaces,* 133:191 (1988).
27. Bottero, J. Y., Tchoubard, D., Axelos, M. A. V., Quienne, P., and Fiessinger, F., *Langmuir,* 6:596 (1990).
28. Hess, P. L., Prisch, H. L., and Klein, R. Z., *Physical Review B: Condensed Matter,* 1986, 64.
29. Meakin, P., Donn, B., and Mulholland, G., *Langmuir,* 5:810 (1989).
30. Meakin, P., Zhong-Ying, C., and Deutch, J. M. *J. Chem. Phys.,* 82(8):3786 (1985).
31. Hess, P. L., Frisch, H. L., and Klein, R. *Z. Phys. B,* 64:65 (1986).
32. Meakin, P., Donn, B., and Mulholland, G. *Langmuir,* 5:810 (1989).
33. Poirier, J. E., Thèse de Doctorat de 3° cycle, INPL, Nancy, France, (1979).
34. Glatter, O. and Kratky, O. *Small Angle X-ray Scattering* O. Glatter and O. Kratky, Eds. (New York: Academic Press, Inc., 1982).
35. Gregory, J. *J. Colloid Interface Sci.,* 42:448 (1973).
36. Wang, T. K. and Audebert, R. J. *J. Colloid Interface Sci.,* 119(2):459 (1987).
37. Hahn, H. H. and Stumm, W. J. *J. Colloid Interface Sci.,* 28(1):132 (1968).
38. Mandelbrot, B. B. *Fractals, Form, Chance, and Dimension* (San Francisco: W. H. Freeman & Company Publishers, 1977).

39. Klimpel, R. C. and Hogg, R. J. *J. Colloid Interface Sci.* 113(1):121 (1986).
40. Hogg, R. and Ray, T. In: *Interfacial Phenoma in Biotechnology and Materials Processing,* Y. A. Attia, B. M. Moudgil, and S. Chander, Eds. (New York: Elsevier Publishers, 1988), pp. 543-554.
41. Pinnavaia, T. J., Rainey, V., Ming-Shintzong, and White, J. M. *J. Mol. Cat.,* 27(21):3 (1984).
42. Bottero, J. Y., Bruant, M., and Cases, J. M. *Clay Miner.,* 23:213 (1988).
43. Changui, C. H. Ph.D. Thesis, Université Catholique de Louvain-la-Neuve, Belgique (1988).
44. Plee, D., Borg, F., Gatineau, L., and Fripiat, J. J. *J. Am. Chem. Soc.,* 107:2362 (1985).

CHAPTER 7

Drinking Water Treatment by Hydrosoluble Polymers: Mechanisms of Coagulation and Efficiency

T. K. Wang, G. Durand, F. Lafuma, and R. Audebert

A major step in the preparation of drinking water is the elimination (coagulation or flocculation) of colloid particles in suspension in natural water. Besides mineral coagulants, based upon multivalent cationic salts, many polymers were progressively accepted for this purification by the health authorities of industrial countries (United States, Great Britain, Belgium, and more recently France). The recommended polymers are mainly polyelectrolytes or copolymers with both neutral and ionic units. But there are only a few data on their flocculation mechanism and especially on the correlation between structural properties of polymers (molecular weight, ionic unit content for the copolymers) and practical behavior in water treatment. Information about the amount of polymer needed to obtain an optimal flocculation, residual turbidity of the supernatant, rate of sedimentation, and rheological properties of the flocs is needed.

We used synthetic water: model suspensions were obtained by adding known mineral (clay or precipitated silica) to deionized water (milli-Q system of Millipore). By treating these suspensions by a series of cationic copolymers of acrylamide we showed that, depending on molecular weight and composition of the copolymers, two types of flocculation mechanisms are involved which govern the floc behavior.

MATERIALS

Polymers

Copolymers were prepared, as described previously,[1] by free radical polymerization of commercially available monomers: acrylamide (AM) and N,N,N,-trimethylaminoethyl chloride acrylate (CMA):

$$--(-CH_2-CH-)----(CH_2-CH)---$$

```
         |                |
        C=O              C=O           Cl⊖CH₃
         |                |             ╱
        NH₂           O-CH₂-CH₂-N—CH₃
                                   ╲
                                  ⊕ CH₃
        AM              CMA
```

According to the reactivity ratio of the two monomers involved[2] and their sequence distribution in the chain studied by NMR[3] the copolymerization reaction is approaching equilibrium. The chemical composition of these copolymers is expressed by the ionicity parameter τ with $\tau = 100\ n/(n + m)$ and where n and m are, respectively, the number of CMA and AM units in the chains. The complete range of cationicity was covered ($\tau = 0$ to 100%), and a series of samples of various molecular weight was obtained in the range 8×10^4 to 3×10^6. The detailed polymer characterization was given previously.[1]

Clay

We used a Wyoming montmorillonite from Ward's Natural Science which was purified and equilibrated by NaCl[4] to prepare sodium montmorillonite suspensions (0.05 to 0.5 g L^{-1}). The suspended particles are platelike (200 nm × 200 nm × 1 nm) with a high specific area of 800 m² g^{-1} and a great cation exchange capacity (95 meq/100 g).[5] For high solid concentrations, platelets are stacked into tactoids,[6] but the aggregation tendency decreases with the clay concentration, and taking into account various observations including low-angle X-ray diffraction or diffusion,[6,7] we think that platelets have a behavior of independent particles for our low concentration regimes.

Silica

We used precipitated silicas. Most of the results are obtained with "syton W30" supplied by Monsanto. From electron micrographs, particles appear almost spherical with diameters between 20 and 200 nm, leading to a mean specific area of about 32 m² g⁻¹. A similar description of this material is given elsewhere[8] and recent BET measurements confirm the specific surface area.[9] For sedimentation study purpose, monodisperse beads of 1-μm diameter, kindly supplied by Rhône Poulenc, were used.

TECHNIQUES

Adsorption Isotherms

Total organic carbon was titrated in the supernatant by means of a Dohrmann DC 80 analyzer after 2 hr standing and 1 hr 15,000-g centrifugation (Cantow Instruments Model 3229).

Flocculation Tests

They were performed according to the normalized jar test with controlled agitation (3 min rapid mixing at 100 rpm, 20 min slower mixing at 40 rpm) at room temperature. After 30 min standing, the extent of clarification was estimated from the light scattered with a FICA 40,000 apparatus equipped with a He-Ne laser source. The electrokinetic potential was measured with a zetameter (laser Zee 5600 Pen Kem Inc), and the rate of settling was obtained simply from the time required by the flocs to fall down between two marks on the jar.

Layer Thickness Measurements

Hydrodynamic layer thickness was obtained by comparison of relative viscosities of silica suspensions before and after polymer adsorption. The principle and the experimental conditions of the technique are given elsewhere.[10] The simple measurement of sedimentation velocity of 1-μm monodisperse silica beads, again before and after polymer adsorption, also simply leads to the hydrodynamic layer thickness.[10]

RESULTS AND DISCUSSION

Polymer-Mineral Surface Affinity

Our experiments are performed at pH around 6.5 (clay) or 7.5 (silica). In these conditions both silica and clay are negatively charged (except, perhaps,

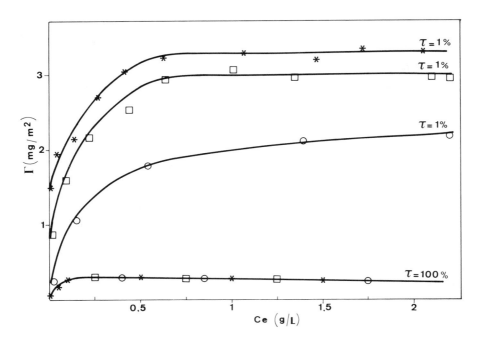

FIGURE 1. Influence of the molecular weight ($\overline{M}_w$) adsorption isotherms for copolymers with $\tau = 1\%$ and $\tau = 100\%$ on Na-montmorillonite 0.5 g/L, $C_{NaCl} = 1.2 \times 10^{-3}$ M, pH = 6.5. $\overline{M}_w = 2 \times 10^6$ (*), 2×10^5 (□), and 8×10^4 (○). (From Reference 13 with permission of Academic Press.)

the lateral side of montmorillonite platelets). Zeta potential (ζ) is -40 mV for clay and -50 mV for silica (Syton type) suspensions. So the cationic units of the copolymers are very strongly adsorbed (the enthalpy change is of 4 kT per cationic moiety adsorbed on clay).[9] As for acrylamide units we found, like Bruhant et al.,[11] that they are adsorbed on clay, with a comparatively small enthalpic change, but they are repelled by a precipitated silica surface. Adsorption kinetic measurements show that the time needed for the achievement of adsorption increases from some minutes for highly cationic flocculants to some hours for almost neutral copolymers.

Adsorption Isotherms

For the two mineral surfaces, the polymer molecular weight effect on the adsorption isotherm profile strongly depends on the cationicity (Figures 1 and 2). For $\tau > 15\%$, high affinity-type adsorption isotherms are obtained; the maximum amount (Γ_{max}) of polymer adsorbed per surface area unit is rather smaller than 1 mg/m² and practically independent of the molecular weight.[12-14]

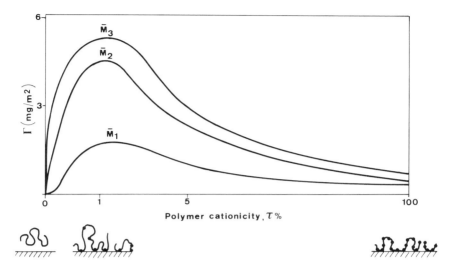

FIGURE 2. Effect of copolymer cationicity on the maximum amount of polymer (Γ_{max}) adsorbed on silica suspensions for different molecular weight ($M_1 = 10^5$, $M_2 = 5 \times 10^5$, $M_3 = 3 \times 10^6$). (For a convenient expansion of the cationicity scale we plotted log $(1 + 10^{-2} \tau)$ in abscissa.) Expected configurations in each cationicity range are suggested. (From Reference 10 with permission of Academic Press.)

This result can be understood by considering that electrostatic interactions are the driving forces of the adsorption and that long-range repulsions between ionized polymer loops prevent a dense polymer packing on the mineral surface. This result agrees with theories of polyelectrolyte adsorption[15,16] and experimental results on the ionic strength effect.[17]

The picture is completely different for copolymers with $\tau < 15\%$. As predicted for neutral polymers in good solvents,[18,19] Γ_{max} increases with the polymer molecular weight and tends to level off for very long chains. For such a kind of neutral or practically neutral polymers, the amount of polymer adsorbed is limited by steric hindrances in the trains-loops-tails system of the adsorbed chain, which leads to a Γ_{max} value of about 1 to 2 mg/m² usually observed for all types of adsorbed flexible macromolecular chains. However, for both clay and silica, an overcrowding effect is observed when the copolymer contains a small ratio (1 to 5%) of the strongly attracted cationic units. This result is expected on the basis of a recent theory of statistical copolymer adsorption[20] for which the polymer configuration is depicted under Figure 2 and where the loops are stretched in the surrounding solution (at the same time, a large layer thickness is predicted).

Layer Thickness Measurements

Effectively (Figure 3) for a copolymer with low τ value (1%), an unusual hydrodynamic layer thickness, δ_H, was observed (δ_H is thicker than 100 nm for

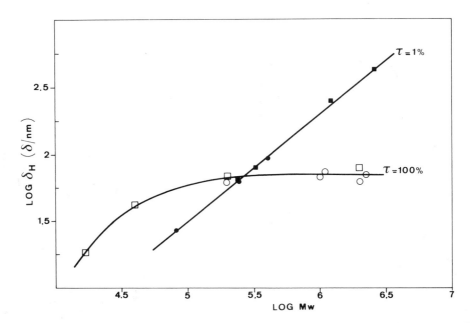

FIGURE 3. Effect of polymer molecular weight $\overline{M}_w$ on hydrodynamic layer thickness δ_H determined by viscometry (squares) and sedimentation (circles) measurements. Open symbol: $\tau = 100\%$, closed symbols: $\tau = 1\%$. (From Reference 10 with permission of Academic Press.)

$M_w = 10^6$ since in similar conditions the measured thickness for a polyethylene oxide does not exceed 20 nm). The layer thickness increases with M_w according to a power law as expected in various theories.[21,22]

For $\tau = 100\%$, the behavior is very different; since for a large range of molecular weight, the layer thickness practically does not depend on M_w ($\delta_H = 70$ nm). This result can be understood since small macromolecular loops are not stable, due to strong electrostatic repulsions between vicinal chain units. So δ_H is probably governed by a critical loop size, practically independent of the molecular weight except for very low M_w.

Flocculation Behavior

We studied the evolution of the zeta potential and the turbidity of the supernatant by adding increasing amounts of polymers (Figure 4). We found again that ionicity and molecular weight are the two basic parameters.

For highly cationic copolymers, the progressive addition of flocculant reverses the zeta potential of the particles. The optimum flocculation concentration (ofc) is obtained for a global neutrality of the particles ($\zeta = 0$ practically coincides with the minimum of the supernatant turbidity). This neutrality is obtained for

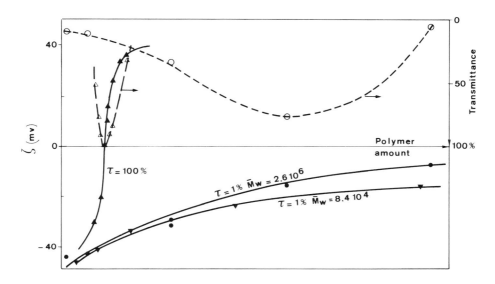

FIGURE 4. Evolution of the zeta potential (mV, filled symbols) and the transmission (τ %, empty symbols) vs the concentration of the polymer introduced C_p (g/L). Silica (syton W 30) concentration 4.6 g/L in deionized water. $\tau = 100\%$ (result independent of the polymer molecular weight); $\tau = 1\%$, $\overline{M}_w = 2.7 \times 10^6$; $\tau = 1\%$, $\overline{M}_w = 8.4 \times 10^4$ (no flocculation occurs). (From Reference 13 with permission of Academic Press.)

a given amount of polymer (i.e., cationic units), whatever the flocculant molecular weight is, polymers are strongly adsorbed on the particles according to the patchwork model proposed by Gregory (Figure 5).[23] Such a behavior is consistent with the major role of electrostatic effects in this flocculation mechanism. When the amount of polymer adsorbed is large enough to give a global charge balance, the long-range interaction between particles is attractive (Van der Waals forces) and the short-range forces depend on the sign of the faced areas of the particles and it can be strongly repulsive (Figures 5a or 5b) or strongly attractive (Figure 5c). Thanks to the strongly attractive situation, a good shear-resistant floc is finally obtained. Both electron micrography[24] and neutron scattering measurement[25] indicate an open fractal structure.

For copolymers of poor ionicities, the zeta potential may stay negative even by adding an excess of polymer. Then only high molecular weight polymers are effective as flocculants (Figure 4) and the ofc decreases with the polymer molecular weight. This suggests a bridging mechanism (Figure 6). For a neutral or almost neutral polymer, the original electrostatic repulsion between particles is not drastically modified by the polymer adsorption (its range, which can be referred to as the Debye length, δ_D, is schematized by the dotted line around the particle). For a low molecular weight polymer, its layer thickness δ_H is always smaller than δ_D and the particle-particle interaction is always repulsive whatever the polymer amount is. When $\delta_H \gg \delta_D$, i.e., by using a polymer of

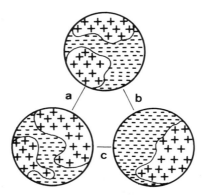

FIGURE 5. Short range interactions between particles during a flocculation process. Particles are assumed to be originally negatively charged; they are examined here after the adsorption of some large cationic macromolecules: interactions a and b are repulsive, c attractive. (From *J. Chim. Phys.* 87:1859 (1990). With permission of Elsevier.)

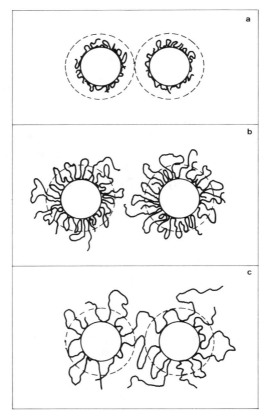

FIGURE 6. Influence of molecular weight and amount of polymer adsorbed on the bridging ability. Dotted circles schematize the range of the electrostatic repulsion. a and b: no flocculation; c: bridging. (From *J. Chim. Phys.* 87:1859 (1990). With permission of Elsevier.)

high molecular weight, the loops or tails of an adsorbed chain can be large enough to "capture" another particle in spite of the electrostatic repulsion. This bridging effect can be effective only if the surface of each particle is not saturated by the polymer (this means that an excess of polymer induces a restabilization). Even if the attractive bridging situation is observed, the approach between the two bridged particles is prevented by the short-range electrostatic repulsions increased by steric chain repulsion. So particles are not in contact in the floc. This result was confirmed by neutron scattering measurements.[25] With very high molecular weight bridging polymers, the mechanical properties of the flocs obtained are compatible with further filtration processes. However, for intermediate molecular weights, the particle-particle interaction is scarcely attractive; on a local scale, the floc has a liquid-like structure and macroscopically the floc is a sticky mud.

CONCLUSION

The flocculation of charged colloidal suspensions by polyelectrolytes of opposite signs is governed by ionicity and molecular weight of the polymers. For highly ionic flocculants, electrostatic interactions play a major role. The ofc is observed practically when ζ is close to zero and does not depend on the polymer molecular weight. Large shear-resistant flocs are obtained. By progressively decreasing the ionic units in the copolymer, below a ratio of about 15%, a completely different aggregation mechanism due to long-range interparticle bridging occurs. In this situation, the ofc decreases with the polymer molecular weight and only very high M_w might lead to interesting systems regarding the amount of needed polymer and the mechanical properties of the flocs.

ACKNOWLEDGMENT

We are greatly indebted to Lyonnaise des Eaux/Dumez and to the GS "Traitement Chimique des Eaux" who supported a part of this work.

REFERENCES

1. Mabire, F., Audebert, R., and Quivoron, C. *Polymer* 25:1317 (1984).
2. Tanaka, H. *J. Polym. Sci. Polym. Chem. Ed.* 24:29 (1986).
3. Lafuma, F. and Durand, G. *Polym. Bull.* 21:315 (1989).
4. Poinsignon, C. Thèse, Nancy (1977).
5. Caillere, S., Henin, S., and Rautureau, M. In: *Minéralogie des Argiles* Masson, S. A. Ed., (Paris Tome 1:INRA Publ., 1982).
6. Fripiat, J., Cases, J., Francois, M., and Letellier, M. *J. Colloid Interface Sci.* 89:378 (1982).
7. Fukushima, J. *Clays Clay Min.* 32:320 (1984).

8. Dhadwal, H. S. and Ross, D. A. *J. Colloid Interface Sci.* 76:478 (1980).
9. Denoyel, R. et al. *J. Colloid Interface Sci.* 139:281 (1990).
10. Wang, T. K. and Audebert, R. *J. Colloid Interface Sci.* 121:32 (1988).
11. Bruhant, M. Thèse, Nancy (1986).
12. Durand-Piana, G., Lafuma, F., and Audebert, R. *J. Colloid Interface Sci.* 119:474 (1987).
13. Wang, T. K. and Audebert, R. *J. Colloid Interface Sci.* 119:459 (1987).
14. Wang, T. K., Piana, G., Lafuma, F., and Audebert, R. *Flocculation in Biotechnology and Separation Systems* (Amsterdam: Elsevier Science, 1987), pp. 165-173.
15. Van der Schee, H. A. and Lyklema, J. *J. Phys. Chem.* 88:6661 (1984).
16. Papenhuijzen, J., Van der Schee, H. A., and Fleer, G. J. *J. Colloid Interface Sci.* 104:540 (1985).
17. Durand, G., Lafuma, F., and Audebert, R. *Prog. Colloid Polym. Sci.* 266:278 (1988).
18. Scheutjens, J. and Fleer, G. J. *J. Phys. Chem.* 83:1619 (1979).
19. Bouchard, E., Auvray, L., Cotton, J. P., Daoud, M., Farnoux, B., and Jannink, G. *Prog. Surf. Sci.* 27:5 (1988).
20. Marques, C. M., and Joanny, J. F. *Macromolecules* 23:268 (1990).
21. Cohen Stuart, M., Waajen, F., Cosgrove, T., Vincent, B., and Crowley, T. L. *Macromolecules* 17:1825 (1984).
22. De Gennes, P. G. *Macromolecules* 14:1637 (1981).
23. Gregory, J. *J. Colloid Interface Sci.* 42:448 (1973); 55:35 (1976).
24. Wang, T. K., Thesis, Paris (1987). (Experiments in collaboration with Pr FAVARD Lab. Cytologie Experimentale CNRS).
25. Cabane, B., Wong, K., Wang, T. K., Lafuma, F., and Duplessix, R. *Colloid Polym. Sci.* 266:101 (1988).

CHAPTER 8

The Role of Colloids and Dissolved Organics in Membrane and Depth Filtration

M. R. Wiesner

SUMMARY

The performance of both depth filtration and membrane filtration units is affected by the concentration and nature of fine colloidal and macromolecular material in the raw water. Colloid size affects the performance of depth filtration units through its role in particle transport and removal within a filter. In contrast, the removal of materials larger than the effective cutoff of membranes is largely independent of particle size. Nonetheless, particle size plays an important role in determining the development of pressure drop and flux across membrane filtration units. The concentration of dissolved organic materials may be the determining factor in selecting an appropriate filtration technology and pretreatment scheme for liquid-solid separation. High concentrations of dissolved organic carbon (DOC) may eliminate direct filtration as a treatment option due to the volume of particulate material created after coagulant addition. In membrane filtration, high concentrations of DOC may irreversibly foul the interior matrix of the membrane and deposit on the membrane surface. Particles that are most difficult to remove in depth filtration are most likely to reduce the flux across pressure-driven membrane filters. As in depth filtration, conditioning the particle size distribution before membrane filtration may improve process performance.

INTRODUCTION

Despite many apparent differences between depth and membrane filters, these unit processes share several similarities with regard to the processes that control mass transport and process efficiency. This paper discusses some of the impacts of colloidal and macromolecular materials in water on the performance of depth and membrane filters.

DEPTH FILTRATION, COLLOIDS, AND DISSOLVED ORGANIC MATERIALS

Considerable work has been done on the relation between particle size and performance in depth filtration. Performance is typically evaluated in terms of particle removal (quality) and headloss development (quantity). Mechanistically, particle removal has been conceptualized as a two-step process of transport up to followed by attachment to the filter media. The development of models for the transport step begins with a force balance on a particle moving through the filter bed. A particle following a streamline within the filter bed may exit its streamline and come in contact with a filter grain as the result of Brownian or gravitational forces. In addition, London-van der Waals and electrical double-layer forces are likely to be important at small separation distances between the particle and the media and must be considered in modeling the particle attachment step. In the absence of all forces other than convective fluid motion, contact between suspended particles and the filter media may still occur. Interception of particles by the filter media may occur if the radius of the particle is equal to or greater than the distance from the streamline to the surface of the filter media.

Expressions for particle transport by interception, gravity, and Brownian diffusion can be combined to calculate particle removal in deep-bed filters.[1] The efficiency, η_B, of a spherical filter grain or "collector" of size d_c in removing particles of size d_p, by Brownian diffusion has been calculated by Levich[2] as:

$$\eta_B = 0.9(kT/\mu d_p d_c v_o)^{2/3} \qquad (1)$$

where k is Boltzmann's constant
T is the absolute temperature
μ is the absolute viscosity of the fluid
v_o is the approach velocity (flow divided by filter plan area)

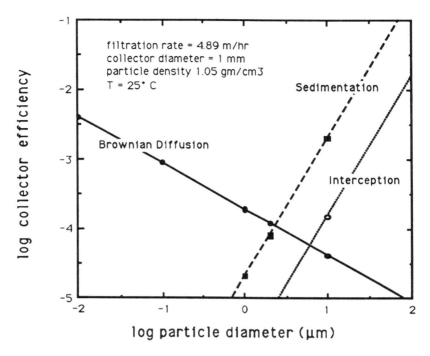

FIGURE 1. Collector efficiency of a spherical filter grain as a function of particle diameter. (After Yao, Habibian, and O'Melia.[1])

The collector efficiencies due to interception (η_I) and gravity (η_G) have been presented by Yao[3] as:

$$\eta_I = (3/2)(d_p/d_c)^2 \tag{2}$$

$$\eta_G = (\rho_p - \rho)gd_p^2/(18\mu v_o) \tag{3}$$

where ρ_p is the particle density
ρ is the density of the fluid
g is the acceleration of gravity

Yao, Habibian, and O'Melia[1] calculated the collector efficiency, η_o, of a filter grain as the sum of Equations 1 through 3. Their calculations indicate that a minimum in the removal efficiency occurs at particle diameters near 1 μm for conditions that are typically encountered in water filtration (Figure 1).

Additional forces and complexities of the fluid flow have been included in models of particle deposition in deep-bed filters. Several researchers have considered the effect of neighboring filter grains on the collector efficiency of a single sphere as described by the Happel sphere-in-cell model.[1,4,5] Rajagopalan and Tien[6,7] incorporated London and double-layer forces with hydrodynamic retardation corrections into their trajectory analysis. Hydrodynamic retardation is significant at small separation distances between moving particles and a fixed boundary such as a the surface of a grain of filter media. The effect is manifested as a deviation from the Stokes expression for the drag force on the suspended particle. Rajagopalan and Tien[5] present the following close-form expression for the collector efficiency, η_o, that includes the effects of hydrodynamic retardation:

$$\eta_o = 4A_s^{1/3}N_{Pe}^{-2/3} + A_s N_{Lo}^{1/8} N_R^{15/8} + 3.38 \times 10^{-3} A_s N_G^{1.2} N_R^{-0.4} \qquad (4)$$

where A_s incorporates corrections from the Happel sphere-in-cell model and is dependent on filter bed porosity
N_{Pe} is a dimensionless Peclet number calculated from the filtration rate, grain size, and particle diffusion coefficient
N_{Lo} expresses the relative importance of London forces and is proportional to the Hamaker constant for the particles
N_R is the ratio of this particle to filter media diameter
N_G captures the effect of gravity as a transport mechanism

In comparison with the sum of Equations 1 to 3, Equation 4 produces a deeper minimum in the efficiency of filter grains in capturing suspended particles near 1 μm in diameter.

These considerations do not indicate whether or not a particle that is transported to the surface of a filter grain will actually attach to the filter grain. The parameter "α" is often introduced to modify the collector efficiency, η_o, by the fraction of transported particles that are actually removed from the suspension such that the apparent collector efficiency is given as:

$$\eta_r = \alpha \eta_o \qquad (5)$$

The collision efficiency factor, α, is considered to be a function of surface interactions between the particle and the collector and is affected therefore by the solution chemistry. In most water treatment applications, coagulants are added to water to decrease particle stability and enhance the attachment of particles. The value of α is assumed to be close to one in this case and particle deposition is favorable. Equation 5 can be incorporated into a mass balance on a depth filter, and the resulting differential equations can be solved to describe the concentration of particles remaining in suspension as a function of filter depth.

Particles retained in a filter become collection sites for subsequent particle removal. This increase in collector area within the filter produces an increase in removal efficiency and headloss across the filter. O'Melia and Ali[8] found good agreement between calculations based on a revised version of this model that considers the effect that particles deposited over time in the filter may have on the removal of other particles. The model describes the initial period of filter ripening and headloss development during the filtration of a monodisperse suspension of particles. The development of headloss in this model is described using a modified Kozeny equation. The Kozeny equation calculates headloss across a clean filter bed as a function of the surface area of the filter media in the bed; greater surface areas produce more headloss. By augmenting the surface area of the filter media to include particles deposited in the filter, the headloss across the filter, h_f, can be calculated at any time:

$$h_f = 4.5 \, L\nu(v_o/g)\{(1 - f)^2/f^3\}\{A_c + \beta A_p)/(V_c + V_p)\}^2 \qquad (6)$$

where ν is the kinematic viscosity of the fluid
A_c and A_p are the surface areas of the filter media and deposited particles respectively per unit volume of filter bed
V_c and V_p are the volumes of the filter media and deposited particles, respectively, per unit volume of filter bed
β is the fraction of surface area of removed particles that contributes to the headloss

The ratio of particle surface area to volume decreases with increasing particle diameter. Thus, for an equal mass of deposit, large particles should produce less headloss than small particles.

The configurations of several operating water treatment plants and the raw waters they treat have been compared with configurations calculated to be cost-optimal for a given particle size distribution in the raw water.[9] Cost-optimality was determined by constrained optimization of a cost function for filtration with a model for filter ripening. Significant deviations of plant design from the predicted least-cost configuration were noted for raw waters containing high concentrations of dissolved organic carbon (DOC). This is anticipated in light of the significant increase in particle volume concentration that may occur when waters containing DOC are flocculated. As an approximation it has been suggested that at a minimum, 1.6 ppm of particle volume per milligram of DOC per liter should be added to the measured particle volume concentration in the raw water when estimating the final particle volume loading to be treated after coagulant addition. On this basis, direct filtration and contact filtration would not be expected to be cost-efficient treatment configurations for raw waters containing more than roughly 6 mg/L of DOC. Significant concentrations of particulate materials in the raw water may reduce this limit. The increase in particle volume concentration that occurs when waters containing appreciable DOC are flocculated is due primarily to two factors. First, higher concentrations of DOC in the raw water necessitate higher coagulant doses which in turn increase the solids loading to the filter. The need for higher coagulant doses arises from the stabilizing effect that organic materials in natural waters have on colloidal materials. Charged functional groups on adsorbed organic compounds impart stability to colloidal materials and may interact with coagulant species to create a coagulant demand. By satisfying the coagulant demand of the organic materials, colloids are more easily attached to filter media and to each other. Second, interactions between DOC and coagulant species result in a higher concentration of organic material associated with the solid phase. Precipitation and/or adsorption of DOC to hydroxide flocs appears to be the most significant addition to particle volume resulting from coagulant addition.

MEMBRANE FILTRATION, FLUX, AND FOULING

Analogous to depth filtration, performance criteria for membrane filtration include both quality (particle removal) and quantity (permeate flux). Materials with an effective size that is larger than the minimum pore size (or membrane cutoff) will be removed by the membrane to a first approximation. Thus, if the pore size distribution is known, the "clean filter" removal of particles can be estimated for a membrane filter. Permeate flux is closely related to the pressure drop across the membrane. Materials that accumulate on the membrane surface produce an additional layer of resistance to flow that is manifested as an increase in the pressure drop (or headloss) across the membrane. The size of particles deposited on the membrane will affect the hydraulic resistance of this layer. For a unit depth of deposit, larger particles should produce a layer of lower resistance than smaller particles. This is illustrated in Figure 2 where calculations of flux across an ultrafiltration membrane with a cake of particles deposited on the surface is plotted as a function of cake thickness for particles of different sizes. For conditions typical of ultrafiltration, particles larger than 1 μm are calculated to have very little effect on permeate flux, even when relatively thick cakes are assumed to be deposited on the membrane.

Particle size also plays a role in determining the net deposition of particles on membranes and several mechanisms have been proposed by which particles may be transported away from the surface of membranes. As particles concentrate near the surface of the membrane, they may diffuse back into the bulk flow. By the Stokes-Einstein equation for the particle diffusion coefficient, the diffusive

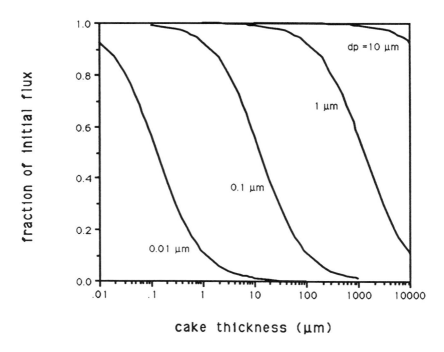

FIGURE 2. Permeate flux (as a fraction of clean membrane flux) across an ultrafiltration membrane ($d_{pore} = 0.01$ μm) as a function of the thickness of colloidal deposits on the membrane surface. Calculations are shown for deposits formed from particles of varying size. For an equal thickness of deposited particles, smaller particles reduce permeate flux more than large particles.

flux of particles away from the membrane is predicted to increase with decreasing particle diameter. Thus, the diffusive flux of small particles from the membrane surface should be greater than that of larger particles.

Green and Belfort[10] extended the earlier work of Cox and Brenner[11] to describe the effect of lift forces on rotating particles as a mechanism of back-transport in membrane filtration. Lateral migration is predicted to increase with increasing particle diameter since the inertial lift force is proportional to the particle diameter cubed. Several models for back-transport have been presented that consider shear-induced transport or scouring of particles deposited on the membrane.[12-16] Although the details of these models may differ, they are all fundamentally based on the concept that the shear force of fluid moving across the membrane initiates particle migration and at a critical point overcomes the advective force of fluid moving through the membrane (as well as any gravity force or interparticle attraction).

Order-of-magnitude estimates of mass transfer by Brownian diffusion, v_B, inertial lift, v_L, and shear-induced diffusion, v_S, can be calculated as a function of particle size:[17]

$$v_B = kT/(3\pi\mu R d_p) \qquad (7)$$

$$v_L = u_o^2 d_p^3/(32\nu R^2) \qquad (8)$$

$$v_S = 0.05\, u_o d_p^2/(4R^2) \qquad (9)$$

where d_p is the particle diameter
 μ is the absolute viscosity
 k is Boltzmann's coefficient
 T is the absolute temperature
 R is the radius of a hollow fiber or capillary membrane (or equivalently the distance to the centerline between parallel porous walls)
 u_o is the centerline maximum velocity in the hollow fiber
 ν is the kinematic viscosity

For conditions typical of a hollow fiber membrane, a minimum in the back-transport rate of particles from the membrane is calculated to occur at a particle diameter of approximately 0.2 μm (Figure 3). For particles smaller than 0.2 μm, the theoretical back-transport rate is dominated by Brownian diffusion. Larger particles are transported away from the membrane by inertial lift and shear-induced diffusion. The back-transport velocities shown in Figure 3 can be compared with the permeate fluxes in hollow fiber membranes to estimate the conditions in which back-transport will offset advective deposition on the membrane. Typical ultrafiltration fluxes are on the order of $10^{-2.5}$ cm/sec. Particles larger than 3 μm have a back-transport velocity due to lateral migration that is greater than the permeation velocity. Thus, these larger particles are not expected to contribute to the fouling of ultrafiltration membranes. An increase in flux increases the range of particles sizes over which deposition is anticipated. For example, the smallest particle that should not contribute to the fouling of a microfiltration hollow fiber membrane is approximately 18 μm.

As a measure of the scattering of visible light ($\lambda \sim 500$ nm), turbidity will be particularly sensitive to those particles in the raw water that are most likely to deposit on the surface of ultrafiltration membranes; particles from approximately 0.1 to 1 μm in size are transported least effectively from membrane surfaces. This is precisely the range of particles sizes that is predicted to be least effectively removed by depth filtration. Coagulation pretreatment should be effective in aggregating these particles to a size that is continuously removed from the surface of the membrane.

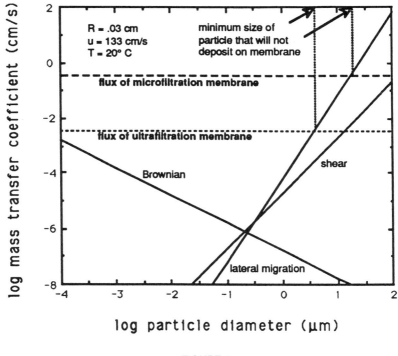

FIGURE 3.

Lahoussine-Turcaud et al.[18] evaluated the fouling characteristics of dispersion composed of materials varying in size over a range typically encountered in water and wastewater treatment applications. They also tested the feasibility of improving membrane performance in water treatment by increasing particle size before membrane filtration using inorganic coagulants.

They found that the membrane fouling behavior of dispersions, made up of particles ranging in size from several nanometers to several hundred microns, was in qualitative agreement with predictions of fouling potential based on particle transport considerations. For the experimental conditions in their study, a minimum in back-transport, and therefore a maximum potential for fouling, is calculated to occur at particle diameters near 0.2 μm. A stable colloidal dispersion of flocculated humic acid composed of 0.2-μm particles produced extensive, yet reversible, fouling of a hollow fiber membrane. These particles are too large to be transported by Brownian diffusion and are too small to be pulled from the surface by shear forces or inertial lift. They are also too large to penetrate into the matrix of an ultrafiltration membrane and thus, most of the reduction in flux can be reversed hydrodynamically. Also consistent with theory, smaller, more mobile materials such as unflocculated humic acid were deposited on the membrane to a lesser extent. As particle size approaches the effective size of membrane pores, fouling due to pore blockage occurs. Experimentally, kaolin (approximately 3 μm in diameter) and coagulated humic and tannic acids (>80 μm) were found to produce little if any fouling.

The characteristics of the DOC in the raw water and the factors affecting its adsorption to membranes may be the most critical factors in determining permeate flux and membrane longevity. In water treatment, naturally occurring organic macromolecules, such as humic and fulvic acids, are likely to foul membranes by adsorbing in pores or on the membrane surface. Deposition in membrane pores produces reductions in membrane flux that are operationally difficult to reverse. Molecules with an effective size that is larger than the size of membrane pores will be rejected at the surface of the membrane and therefore should not be deposited within the matrix of membrane pores. However, much of the DOC in natural waters is of small enough molecular size that materials will penetrate the outer skin of many microfiltration, ultrafiltration, and nanofiltration membranes and adsorb within the membrane matrix. Relative to particle deposition on the membrane surface, DOC may produce large irreversible losses in membrane flux through adsorptive fouling.[18] The concentration of organic carbon in raw water may be a useful indicator of the rate at which such irreversible losses in membrane capacity might occur. Coagulation pretreatment of raw waters should be effective in reducing these losses if the DOC is extensively complexed. If, in addition, the resultant particles are aggregated to form large floc, flux reductions due to surface deposition may also be reduced. However, a pilot test of coagulation pretreatment of Seine River water showed that small molecular weight polysaccharide materials remain in solution after flocculation.[19] The irreversible fouling of the hydrophobic membranes used in this work was attributed to these small molecular weight materials.

CONCLUSIONS

There is a range of particle sizes between approximately 0.1 and 1 μm in which particle transport processes in both granular media depth filtration and membrane filtration are at a minimum. In depth filtration, this range of poor transport results in poor removal of particles from water as it passes through the filter. In membrane filtration, this range of minimum transport is likely to result in poor removal of particles from the membrane surface, resulting in a reduced flux of permeate.

The performance of both depth filtration and membrane filtration units is greatly affected by the presence of dissolved organic materials in the raw water.

The impacts of these dissolved substances on filter performance should be considered in the evaluation and selection of filtration technologies. In depth filtration, dissolved organic materials are likely to necessitate higher coagulant doses in order to destabilize particles and remove them in the filter. Coagulant addition and the association of organic materials with the newly formed solid phase will increase the loading of particles onto filters and this loading must be considered in the filter design. Raw waters that contain less than 6 mg/L DOC and less than 20 mg/L of suspended solids are possible candidates for direct or contact filtration. Membrane filters may be fouled irreversibly by adsorption of organic molecules to membrane surfaces. Irreversible fouling is likely to be greater when the size of organic molecules is less than the molecular weight cutoff of the membrane. Coagulation pretreatment to reduce fouling by dissolved organic materials is only likely to be effective when very high percentages of the organic material are complexed by the coagulant. In this case, measurements of UV absorbance or DOC in a conventional jar test can be used to indicate the effectiveness of coagulation pretreatment.

REFERENCES

1. Yao, K. M., M. T. Habibian, and C. R. O'Melia. "Water and Waste Water Filtration: Concepts and Applications," *Environ. Sci. Technol.* 5(11):1105-1112 (1971).
2. Levich, V. G. *Physicochemical Hydrodynamics* (Englewood Cliffs, NJ: Prentice-Hall, 1962).
3. Yao, K. M. "Influence of Suspended Particle Size on the Transport Aspect of Water Filtration," Ph.D. thesis, University of North Carolina, Chapel Hill, NC (1968).
4. Spielman, L. A., and J. A. Fitzpatrick. "Theory for Particle Collection under London and Gravity Forces," *J. Colloid Interface Sci.* 42:607-623 (1973).
5. Rajagopalan, R., and C. Tien. "Trajectory Analysis of Deep-Bed Filtration with the Sphere-in-Cell Porous Media Model," *Am. Inst. Chem. Eng. J.* 22:246-255 (1976).
6. Rajagopalan, R., and C. Tien. "Single Collector Analysis of Collection Mechanisms in Water Filtration," *Can. J. Chem. Eng.* 55:246-255 (1977).
7. Rajagopalan, R., and C. Tien. "Experimental Analysis of Particle Deposition on Single Collectors," *Can. J. Chem. Eng.* 55:256-264 (1977).
8. O'Melia, C. R., and W. Ali. "The Role of Retained Particles in Deep Bed Filtration," *Prog. Water Tech.* 10(5/6):167-182 (1978).
9. Wiesner, M. R., and P. Mazounie. "Raw Water Characteristics and the Selection of Treatment Configurations for Particle Removal," *J. Am. Water Works Assoc.* 81:80 (1989).
10. Green, G., and G. Belfort. "Fouling of Ultrafiltration Membranes: Lateral Migration and Particle Trajectory Model," *Desalination* 35:129-147 (1980).
11. Cox, R. G., and H. Brenner. "The Lateral Migration of Solid Particles in Poiseuille Flow. I. Theory, *Chem. Eng. Sci.* 23:147-173 (1968).

12. Zydney, A. L., and C. K. Colton. "A Concentration Polarization Model for the Filtrate Flux in Cross-Flow Microfiltration of Particulate Suspensions," *Chem. Eng. Commun.* 47:1-21 (1986).
13. Davis, R. H., and D. T. Leighton. "Shear-Induced Transport of a Particle Layer Along a Porous Wall," *Chem. Eng. Sci.* 42(2):275-281 (1987).
14. LeGuennec, B., and V. Milisic. "Hydrodynamic Model of Cross-Flow Filtration," *Phys. Chem. Hydrobio.* 7(4):183-190 (1986).
15. Fane, A. G. "Ultrafiltration of Suspensions," *Membrane Sci.* 20:249-259 (1984).
16. Leonard, E. F., and C. S. Vassilief, "The Deposition of Rejected Matter in Membrane Separation Processes," *Chem. Eng. Commun.* 30:209-217 (1984).
17. Wiesner, M. R., M. M. Clark, and J. Mallevialle. "Membrane Filtration of Coagulated Suspensions," *J.Environ. Eng.* ASCE 115(1):20-40 (1989).
18. Lahoussine-Turcaud, V., M. R. Wiesner, and J. Y. Bottero. "Fouling in Tangential-Flow Ultrafiltration: The Effect of Colloid Size and Coagulation Pretreatment," *J. Membr. Sci.* 52:173-190 (1990).
19. Lahoussine-Turcaud, V. et al. "Coagulation Pretreatment for Ultrafiltration of a Surface Water," *J. Am. Water Works Assoc.* 82(12):76-81 (1990).

CHAPTER 9

Preliminary Analysis of the Organic Contents in Treated and Raw Water in Guangzhou City

L. Zhicai, C. Wanhua, and Z. Shaojia

Guangzhou Water Supply Company includes nine potable water plants and one industrial water plant. It is an old enterprise with an 80-year history. River water provides raw water for the nine potable water plants except for the Jiangcun Water Plant, where the raw water is a combination of groundwater and river water. Xintung Water Plant is situated on the Donjiang River; Jiangcun Water Plant is on the Liuxi River; and all the other plants are on the Guangzhou section of the Pearl River.

The water in different areas of these rivers have been polluted to different extents by industrial waste water and domestic sewage during the past few years. For example, the annual mean content of ammonia has risen yearly at the Xicuen Water Plant, which supplies about half of the water demand in Guangzhou City. The ammonia level was 0.01 mg/L at the end of the 1950s; 0.2 to 0.3 mg/L at the end of the 1960s; 0.6 to 0.7 mg/L at the end of the 1970s, and is now 0.9 to 1.3 mg/L at the end of the 1980s. The nitrite concentration has also increased.

In order to determine the amount of hazardous organic chemicals in raw and treated water XAD-2 was used as an adsorbent to concentrate samples, and GC/MS with an ion trap and computer were used to monitor the organic materials.

COLLECTION AND PRELIMINARY TREATMENT OF WATER SAMPLES

Resin adsorption was used to concentrate organic material from water (see Figure 1). The sampling time chosen was during the dry period because the

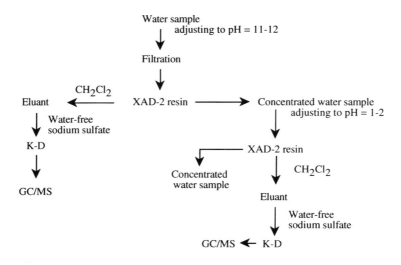

FIGURE 1. Isolation of organic chemical by XAD-2 resin for GC/MS analysis.

water quality is the worst at that time. All samples were obtained from the same location. A 110-L raw water sample is collected in a glass bottle, which is pumped from the intake, 0.3 to 0.5 m below the water surface. An additional 110-L treated water sample was collected according to the mean detention time of the water treatment plant. The pH value of samples are adjusted to 11 to 12 with NaOH. After filtration, the sample is passed through a concentration column filled with XAD-2 resin. Neutral and alkaline organic compounds are concentrated. The water from the XAD-2 resin column is then adjusted to pH 1 to 2 with HCl and passed through another XAD-2 resin column to concentrate acidic compounds. Each resin column is eluted with redistilled dichloromethane. The eluant is collected and mixed with 5 g of anhydrous sodium sulfate and stored overnight. The eluant is then placed in a water bath at 50 to 60°C and purged with nitrogen gas to concentrate the sample up to 1 mL by a Kuderna-Danish evaporative (K-D) concentrator. The sample is stored in a refrigerator.

QUALITATIVE AND QUANTITATIVE ANALYSIS OF ORGANIC CONTAMINANTS

The organic compounds were identified by mass spectrometry and compared with reference standards. The instrument consists of a capillary gas chromatograph and an ion trap detector. An IBM PC/XT was used to process data.

According to the rule of available carbon value, the correct formula weight Fw is calculated from:

$$Fw_i = (7.00 \times M_i)/(100 \times N_c)$$

where Fw_i = correction factor for weight of the component i
        7.00 = available carbon value of the standard *n*-heptane
        M_i = molecular weight of the component i
        100 = molecular weight of the standard *n*-heptane
        N_c = available carbon number of the detected compound

The detected compounds are quantified by extrapolation:

Calculating concentration of the detected component i

$$C_i = (Fw_i \times H_i \times V_s \times C_s)/(H_s \times V_i \times D)$$

where C_i = concentration of component i in µg/L
        Fw_i = correction factor for weight of the component i
        H_i = peak height of the component i in mm
        V_i = intake of the sample in µL
        H_s = peak height of the standard *n*-heptane in mm
        V_s = intake of the standard *n*-heptane in µL
        C_s = concentration of the standard *n*-heptane in µg/L
        D = concentrated factor of the water sample

RESULTS

A total of 210 different organic chemicals were found in the raw and finished water. The chemicals were classified into 19 groups as shown in Table 1. The concentrations of 146 compounds are within the range of 0.01 to 1.63 µg/L. Ten compounds are listed as priority pollutants by the U.S. EPA. They are ethyl benzene; *n*-dibutyl phthalate; *n*-dioctyl phthalate; chloroform; alpha, beta, gamma, and delta BHT; *p,p'*-DDT, and *p,p'*-DDD. Ethyl benzene is only found in the raw water of the Fancun Water plant at a concentration of 0.01 µg/L. Both phthalates, *n*-butyl and *n*-octyl, are plasticizers detected occasionally in raw and

Table 1. Groups and Numbers of Detected Organic Chemicals in Treated and Raw Water

N°	Group	Number	%
1	n-Alkane	19	9.0
2	Alkene	9	4.3
3	Alkyne	1	0.5
4	Arene	9	4.3
5	Nitrile	5	2.4
6	PAH	1	0.5
7	Halohydrocarbon	10	4.8
8	Amine	23	11.0
9	Phenol	2	1.0
10	Alcohol	28	13.3
11	Ether	1	0.5
12	Organic acid	15	7.1
13	Ester	26	12.4
14	Aldehyde	10	4.8
15	Ketone	17	8.1
16	Oxygen heterocyclic ring	10	4.8
17	Nitrogen heterocyclic ring	12	5.7
18	Sulfide	7	3.3
19	Other	5	2.4

Table 2. Annual Mean Content of the Chloroform in Treated and Raw Water in Xicun Water Plant (in µg/L)

Year	1981	1982	1983
Raw water in high tide	1.3	0.5	0.5
Raw water in low tide	1.5	<0.5	<0.5
Treated water	4.7	4.3	4.3

treated water at concentrations of 0.01 to 0.22 µg/L. Maximum allowable values in water are 20 µg/L for ethyl benzene, 200 µg/L for n-dibutyl phthalate, and 1000 µg/L for n-dioctyl phthalate. Chloroform is not found in the raw water, but is found in treated water in the range of 0.02 to 0.21 µg/L. Both chloroform and carbon tetrachloride are higher in treated water. For example, the annual mean content of chloroform in the treated water is two to four times that of raw water during 1981 to 1983 (see Table 2). The chloroform comes from the chloramination of dissolved organic carbon during disinfection. Carbon tetrachloride is a contaminant in the chlorine gas cylinder that is used.

However, according to monthly data, chloroform values are always lower than the China National Health Standard for Drinking Water, which is at 60 µg/L. During the water treatment process, it is important to control the dosage of the chlorine to limit chloroform formation. The pesticides, BHTs, and DDT are also present during monthly monitoring; their concentrations are below the level of environmental concern.

Table 3. Comparison of the Total Peak Areas of Organic Chromatogram

Plant	Type	Peak Area
Xicun	Raw water	9142
	Treated water	5210
Yuancun	Raw water	16240
	Treated water	7290
Shixi	Raw water	13740
	Treated water	8190
Xicun	Sludge of raw water	5.6×10^5
	Flocculating sludge	2.0×10^6

PRELIMINARY EVALUATION OF REMOVAL OF ORGANIC CHEMICALS BY THE TREATMENT FACILITIES

A raw water sludge and a flocculating sludge were collected at the Xicun Water Plant. The result is shown in Table 3. Three conclusions were obtained from the data:

1. A comparison of the total peak areas of organic chemicals in a chromatogram between the raw water and the treated water shows the total amount of organic chemicals are reduced by about 40 to 50% after treatment. This amount may be conservative because research shows a large amount of organic matter is adsorbed by suspended matter in the raw water, which should be removed by filtration during initial water treatment. Moreover, the total peak area of the raw water sludge is 107 times larger than that of treated water, whereas the flocculating sludge is 384 times as large as that in the tap water. Thus, most organic material that is chromatographable can be removed by coagulation with aluminum salts, sedimentation, and filtration.
2. Organic content in water appears to be related to turbidity. Among the three plants, the turbidity of the raw water is the highest in Yuancun and the second is in Shixi. The total peak area of a chromatogram of the raw water is also the largest in Yuancun and it is about two times as much as that in the Xicun Plant.
3. The total peak area of the chromatogram from the flocculating sludge is more than three times larger than in the raw water sludge. Most of the organic matter is removed after flocculation and sedimentation; however, under certain conditions some of the organic matter in the settled sludge may be redistributed in the water. Therefore, sludge must be quickly removed to prevent additional pollution in the water column.

REFERENCES

1. Junk, G. A. et al. *Z. Anal. Chem.* 282:331 (1976).
2. Novak, J. "Quantitative Analysis by Gas Chromatography."

CHAPTER 10

Treatability Evaluation by Simple and Rapid Method

N. Tambo and T. Kamei

INTRODUCTION

Characterizing the water and wastewater with parameters that are meaningful in terms of eventual functioning of the treatment processes is a necessary prerequisite for developing the basic design concepts of water and wastewater treatment systems.

Molecular size distribution patterns have been frequently applied to determine the effect of individual unit processes in a water and wastewater treatment plant. However, as molecular size distribution with respect to only TOC (total organic carbon) or only E260 (ultraviolet absorbance at 260 nm) is not sufficient enough for the complete characterization of water and wastewater, characterization of water and wastewater by use of apparent molecular size distribution with respect to TOC or E260 and TOC/E260 (ratio of TOC to E260) of each fraction of apparent molecular size distribution was carried out and proved as an effective parameter for estimating treated water quality with respect to TOC or E260. Although apparent molecular size distributions have been determined on the basis of TOC and E260 by Sephadex gel chromatography, this chromatographic method is time-consuming and often requires hours for fractionation. Therefore, HPSEC (high performance size exclusion liquid chromatography) using the newly developed GL-W520 column packing (Hitachi Co., molecular weight exclusion limit = 6000) was conducted to characterize the water and wastewater or predict the treatability of coagulation and biological process.

EXPERIMENTALS

Operational Condition for HPSEC

HPSEC was conducted at ambient temperature on a Hitachi 655 HPLC system consisting of a model 655 pump and UV detector operating at 260 nm and a model 833 data module. The column used was a 8 × 350 mm stainless column packed with Hitachi Gelpack GL-W520 (molecular size exclusion limit = 6000) and a 8 × 10 mm precolumn with the same packing material. The void volume (Vo = 7.8 mL) and the total permeation volume (Vo + Vi = 17.2 mL) were determined with Blue Dextran and acetic acid, respectively. Separation of organics was conducted with a phosphate buffer solution (0.02 M KH_2PO_4 + 0.02 M Na_2HPO_4, pH 6.9) at a flow rate of 0.7 mL min^{-1}.

Sample injection volume and fraction volume fractionated using minifraction collector were 0.1 and 0.7 mL, respectively. Recovery of organics applied to the column was determined by comparing TOC of sample for combined fractions passing through the column. Dilution during chromatographic separation necessitated a 20-fold rotary vacuum concentration of samples of low concentration to ensure TOC determination of each fraction by a Shimadzu model 500 TOC analyzer. The relationship between the molecular weight of model compounds and Ve/Vo (ratio of elution volume to void volume) is shown in Figure 1.

Sample Preparation

Some water samples used for HPSEC are shown in Table 1.

Coagulation and Biological Treatment Experiments

Coagulation Experiments

Coagulation experiments were carried out with a multiple stirrer with six paddle mixers. The standard procedure was to add aluminum sulfate (10,000 mg/L stock solution as $Al_2(SO_4)_3$ 18 H_2O) to six 1-L beakers containing water samples first and then adjust the pH with HCl or NaOH. The rapid mixing time (150 rpm) lasted 5 min; slow mixing (30 rpm), 10 min; and settling, 20 min.

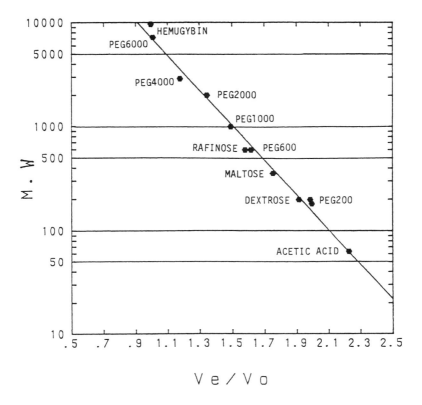

FIGURE 1. Relationship between the molecular weight of model compounds and Ve/Vo for GL-W520 column.

Table 1. TOC and E260 of Water Samples (After 0.45 μm pore size filtration)

	Peat Water Sampled from a Shallow Well Located in a Flat Area	Soil Water Sampled from 5-cm Depth from a Surface of a Well-Forested Hillside	Domestic Wastewater in Sapporo	Filtrate of Heat-Treated Return-Activated Sludge (× 1/40)
TOC	16	77	45	105
E260	0.62	2.48	0.27	0.8
TOC/E260	26	31	171	131

Biological Treatment

Biological treatment was carried out in a 20-L reactor filled with 15 L of water of wastewater samples shown in Table 1. The sample was inoculated with 15 mL supernatant of domestic biological wastewater treatment plant effluent and aerated at 20°C.

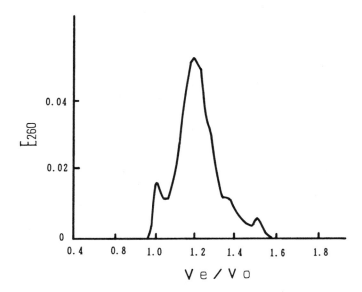

FIGURE 2. Apparent molecular size distribution of peat water.

RESULTS AND DISCUSSION

Column Standardization and TOC Recovery

Column standardization using a solution containing compounds of known molecular weight is presented in Figure 1. Separation of these model organic compounds was good and eluted in a regular order with respect to the molecular weight as shown in Figure 1.

Figure 2 shows the apparent molecular weight distribution of the soluble organics in peat water. Apparent molecular size distribution with respect to E260 can be obtained in 20 min. The average TOC recovery of natural colored organics from the column was almost 100% for several determinations, which suggests that natural colored organics are not irreversibly adsorbed onto the packing material.

Characteristics of Apparent Molecular Size Distribution of Water and Wastewater

It has been known that peat water in a flat area contains a larger fraction of high molecular weight organics compared with stream water in well-forested

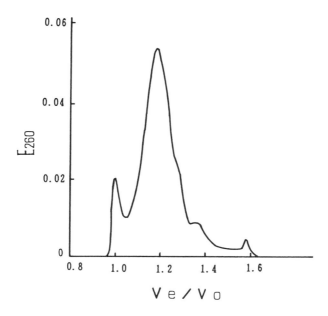

FIGURE 3. Apparent molecular size distribution of hillside soil water (5-cm depth).

hillside areas through various molecular size analysis in the authors' laboratory. Apparent molecular size distribution of soil water sampled at 5 cm depth (Figure 3) from the hillside surface of a forest is quite similar to that of peat water in a flat area (Figure 2). By comparing the apparent molecular size distribution of soil water sampled at various depths with that of river water, it is evident that high molecular weight fractions are adsorbed onto the soils through vertical percolation on run off from the hillside to river (Figure 4).

Domestic Wastewater and Filtrate of Heat-Treated Activated Sludge (200°C 20 kg/cm² for 1 hr)

One of the characteristic features of domestic wastewater and filtrate of heat-treated activated sludge is that the TOC/E260 of lower molecular size fractions such as fraction numbers of 18 and 19 is greater than 400. The TOC/E260 ratio of a water or wastewater of each fraction is indicative to evaluate effectiveness of biological degradation, coagulation, and carbon adsorption process for the fraction. A high TOC/E260 ratio, for example, would indicate that most of the dissolved organic compounds can be degraded biochemically, while a low TOC/E260 ratio would indicate the presence of a significant fraction of biorefractory organic compounds. Although another index such as apparent molecular size

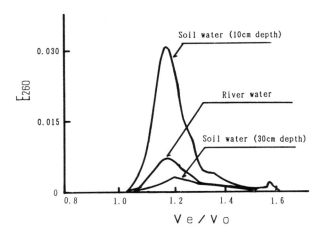

FIGURE 4. Comparison between the apparent molecular size distribution of river water and that of soil water.

distribution is essential for the more accurate treatability evaluation, a high ratio of TOC/E260 would indicate that these kinds of dissolved organic compounds can mostly not be removed by coagulation and carbon adsorption process.

Coagulation Experiments

Figure 5 shows E260 reduction of peat water by aluminium coagulation with settling and membrane filtration. Removal of E260 increased with the increase of aluminium dosage. As the coagulant dose increased, however, a point was reached at which no additional organics were removed, as shown in Figure 5. Removal efficiency with respect to the amount of dosed coagulants was much more effective at pH 5.5 than at 7.0. However, maximum organics removal attainable at pH 7.0 was equal to that of pH 5.5 as shown in Figure 5. It is clear that the percent reduction of organics by coagulation was much greater for the E260-sensitive high molecular weight organics as already reported and shown in Figure 6. This size dependency on the removability has been found for other types of sample waters. Therefore, the relationship between the TOC removal efficiency of each fraction and Ve/Vo for various types of sample water was plotted as shown in Figure 7. As shown, TOC removal efficiency of each fraction of different types of water samples is well correlated with Ve/Vo.

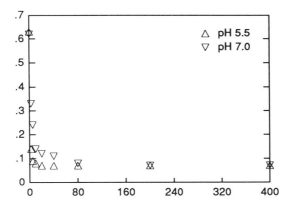

FIGURE 5. Decrease of E260 with increase of alum dosage at pH 5.5 and 7.0.

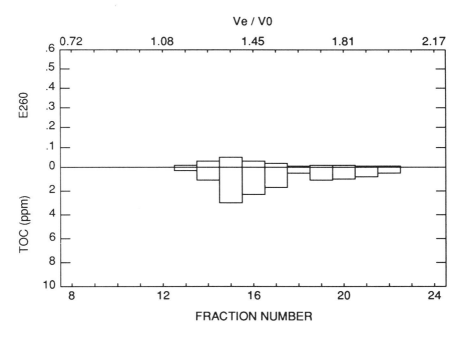

FIGURE 6. Apparent molecular size distribution (with respect to TOC and E260) of peat water after alum coagulation.

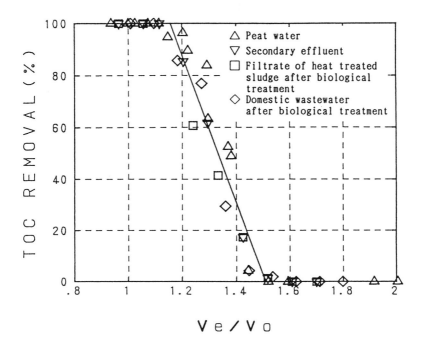

FIGURE 7. Relationship between TOC removal efficiency for water samples (TOC/E260 < 60) and Ve/Vo.

Therefore, removal efficiency of E260-sensitive organics can be formulated on these correlations as follows:

In case of 1.15 < Ve/Vo < 1.50

TOC removal efficiency (%) = 429.2 − 284.4 (Ve/Vo)

In case of Ve/Vo ≤ 1.15 (1)

TOC removal efficiency (%) = 100 (2)

In case of Ve/Vo ≥ 1.50

TOC removal efficiency (5) = 0 (3)

One example of comparison between estimated TOC after coagulation using these equations and the experimental is shown in Table 2. On the other hand it

Table 2. Comparison of Estimated TOC of Each Fraction and Experimental TOC After Coagulation of Peat Water (Maximum Removal)

	TOC (mg/g) Before Coagulation	TOC (mg/g) Estimated	TOC (mg/g) After Coagulation
Peat water	18.0	2.4	2.3

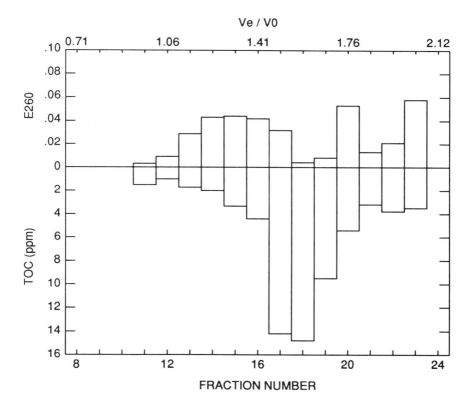

FIGURE 8. Apparent molecular size distribution of domestic wastewater.

is evident by comparing Figures 8 and 9 that only slightly enhanced removal for E260-insensitive organics was obtainable. Therefore, for those water samples containing both E260-sensitive and -insensitive organics, removal efficiency for each fraction can be formulated by the following equation:

$$r_i = \frac{a\text{E260-insensitive TOC} + b\text{E260-insensitive TOC}}{\text{Total TOC}}$$

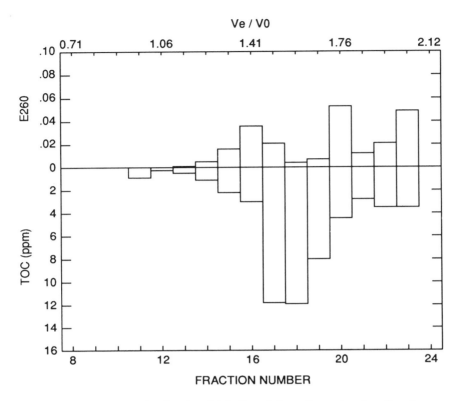

FIGURE 9. Apparent molecular size distribution of domestic wastewater after alum coagulation.

Ratio of E260-sensitive TOC to total TOC (E260-sensitive TOC per total TOC) of each fraction can be estimated by the use of Figure 10, as was shown by Tambo and Kamei.

Biological Treatment Experiments

Figure 11 shows the result of a series of batch-type biological treatments. Result of the variation in the amount of activated sludge (expressed in turbidity) and substrate removal (expressed in TOC) is shown in Figure 11. Apparent molecular size distribution before and after biological treatment is shown in Figures 12 and 13. The highest removal in this biological treatment was obtained for low molecular weight E260-insensitive organics (Figures 12 and 13).

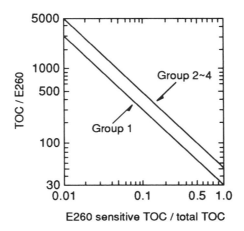

FIGURE 10. Relationship between E260 sensitive TOC/total TOC and TOC/E260 (Ve/Vo < 1.05, Group 1; Ve/Vo > 1.05, Group 2 to 4).

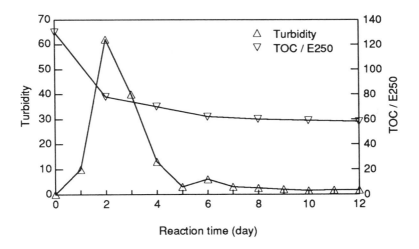

FIGURE 11. Variation in the amount of activated sludge (expressed in turbidity) and TOC/E260.

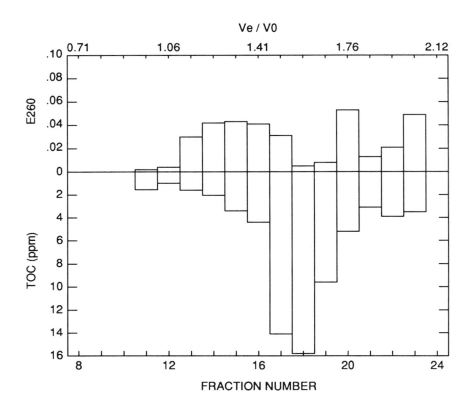

FIGURE 12. Apparent molecular size distribution of domestic wastewater.

The removal efficiency for each fraction by biological treatment can be estimated using the following equation:

$$f(r_i) = \frac{a_i \text{E260-sensitive TOC} + b_i \text{E260-insensitive TOC}}{\text{Total TOC}} \quad (4)$$

where a_i and b_i are removal efficiencies for E260-sensitive TOC and E260-insensitive TOC, respectively

$$a_i = 0.1, \ b_i = 0.80 \text{ for } V_e/V_o \leq 1.4$$

$$a_i = 0.1, \ b_i = 0.98 \text{ for } V_e/V_o \geq 1.4$$

Comparison between the predicted and experimental is shown in Table 3.

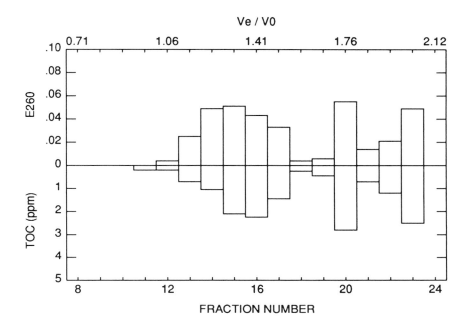

FIGURE 13. Apparent molecular size distribution of domestic wastewater after biological treatment.

Table 3. Comparison between Predicted and Experimental for Biological Treatment

	Raw Water TOC (mg/L)	Predicted TOC (mg/L)	Experimental TOC (mg/L)
Domestic wastewater	45.4	7.2	8.0
Filtrate of heat treated activated sludge	107	28	25

CONCLUDING REMARKS

Apparent molecular size distribution with respect to E260 was obtainable in 20 min with HPSEC. In addition, the number of fractions to be measured by TOC is reduced from approximately 43 fractions obtainable by gravity flow-type Sephadex G-15 gel chromatography (2.5 × 90-cm column with water and 0.01 M NH$_4$OH elution system) to only 13 fractions by this HPSEC.

It was again shown that removal efficiency by coagulation process and biological treatment process is dependant upon two parameters, i.e., apparent molecular size with respect to TOC and TOC/E260. Estimation of water quality after coagulation and biological treatment by use of these two parameters was well correlated with the experimental results.

CHAPTER 11

Automatic Control of the Coagulant Dose in Drinking Water Treatment

C. Hubele

INTRODUCTION

Overview

One of the objectives of potable water utilities is the production of a water with constant quality. To achieve this, the treatment conditions have to be adapted continuously to the raw water quality. Automatization of the plant operation becomes therefore a necessity and explains why during the last years automatic dose control for the clarification process has been studied.

Different methods of control are possible, such as stochastic modeling using parameters which describe the raw water quality[1] or a feedback control based on the raw water quality measured immediately after coagulant injection. However, the latter approach relies heavily on the availability of an adequate sensor, able to give reliable information on the coagulation process. In 1966, the streaming current detector (SCD) was introduced[2] and has since been the object of several publications[3-8] showing the interest of this analyzer. At the same time, most of these studies have used batch experiments and/or synthetic waters and are rather difficult to extrapolate to such complex systems as they are represented by natural surface waters. The purpose of this study was therefore to evaluate the performances of the SCD under practical, yet well-controlled conditions of water treatment. More specifically, the following points were addressed:

- Evaluation of the reliability of the SCD, the reproductivity and the stability of its signal under constant conditions
- Evaluation of the SCD response to different treatment situations, to variations of coagulant dose, pH, and ionic strength
- Development of control algorithms
- Comparison of coagulant doses applied by SCD control with jar test results
- Evaluation of water quality at different control conditions

Presentation of the SCD

The scheme of the streaming current detector cell is shown on Figure 1. It consists essentially of a sampling chamber receiving the flow of the coagulated water (3 L/min, atmospheric pressure). The chamber contains a vertically reciprocating piston which moves inside a cylinder of which one end is closed hermetically. A silver ring electrode is placed at the top and the bottom of the cylinder.

The piston movement acts as a pump and creates a flow of coagulated water between the cylinder and the piston. This causes motion of counter-ions sheared from the colloids which are attached to the surfaces of the piston and cylinder. The movement of these charged particles creates an electrical current which is measured with the electrodes. A presentation of the theoretical principles of streaming current detection has been given by Dentel et al.[7] and is not discussed here.

The sinusoidal current produced is amplified, rectified, and time-smoothed. The final signal obtained is called streaming current, without units, and in general negative when surface raw waters are analyzed, by analogy with zeta potential measurements.

EXPERIMENTAL METHODS

The main part of the experiments was performed on a pilot Pulsator installed at the CEB plant (Compagnie des Eaux de Banlieue de Paris), located at Suresnes, France, a western suburb downstream of Paris. The plant treats Seine river water with a capacity of 53,000 m^3/day and the following treatment train: preozonation, clarification with sedimentation on two Pulsators (4 m/hr), and filtration on rapid sand filters, post ozonation, adsorption on GAC filters, and disinfection by chlorine.

Other experiments were performed at Paray le Monial (East Center of France) on a small coagulation pilot, which consisted of a coagulant injection, in-line

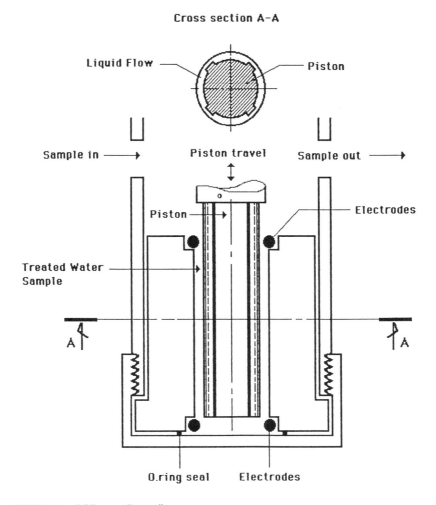

FIGURE 1. SCD sampling cell.

rapid mix and sampling for the SCD. The industrial plant (6000 m³/day) treats water of the Bourbince river by preoxidation, injection of powdered activated carbon (PAC), lime addition, coagulation/flocculation with aluminum sulfate, sedimentation (Pulsator), sand filtration, and disinfection. On this station, the sample was taken after the injection of PAC and aluminum sulfate.

The pilot Pulsator used for the experiments on the Seine river (Figure 2) was operated at a vertical velocity of 4 m/hr. Coagulant was injected into the raw water at a rapid mix chamber (1-min detention time). Between the rapid mix and the vacuum chamber of the Pulsator, flocculation aid could be injected

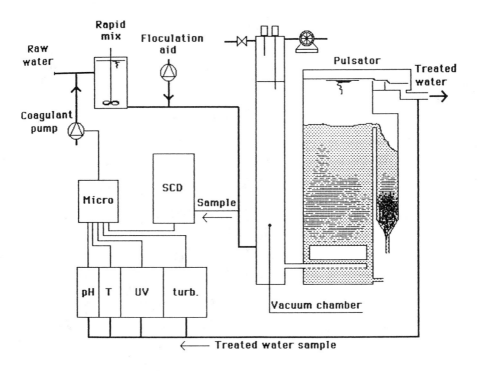

FIGURE 2. Clarification pilot plant.

Table 1. Raw Water Quality of the Seine River (Period 25.01—14.11.88, Seven Samples)

	Minimum	Maximum	Average
Temperature (°C)	6.5	18.0	13
Turbidity (NTU)	3.1	73	22
Resistivity (ohm cm)	1900	2380	2080
pH at 20°C	7.5	8.2	7.8
Alkalinity (as $CaCO_3$) (mg/L)	178	217	202
Calcium (mg/L)	86	100	95
Magnesium (mg/L)	4.4	6.3	5.1
Sulfate (mg/L)	28	43	34
Chloride (mg/L)	15	24	19
TOC (mg/L)	2.8	5.3	3.7
UV 254 nm (L/m)	5.1	12.3	8.8

depending on the nature of the coagulant. The sample for the SCD (Milton Roy, Generation II) was taken downstream of the flocculation aid injection.

Jar tests were used to determine the optimum dose of coagulant at the following experimental conditions: 2 min of rapid mix at 150 rpm, 20 min of flocculation at 30 rpm, and 10 min of sedimentation.

Two types of coagulants were tested: Aqualenc, a commercially available preparation of polymerized $Al_2(SO_4)_3$ corresponding to the formula $Al_2(SO_4)_{3(1-x-y)}(OH)_{6x}Cl_{6y}$, where OH/Al = 1.5 and containing 8.2% in weight of Al_2O_3 and aluminum sulfate (8.2% in Al_2O_3). In the latter case, a naturally derived anionic polymer (sodium alginate) was used as flocculation aid.

RESULTS

Seine River Water

Evaluation of the SCD by Batch Experiments

A summary description of the water quality encountered at the Seine river downstream of Paris is given by Table 1. The data show typical values for a surface water, that is, a relatively high organic pollution and average concentrations of alkalinity and hardness, as well as of neutral salts.

Before starting the evaluation of the SCD response as a function of coagulant dose and raw water quality, the performance and reliability of the analyzer itself was tested. No detailed discussion is presented here; however, satisfying results were obtained. After a change in coagulant dose, the SCD signal was stable again at its new value after about 2 min, mainly due to hydraulic contact time of the sample in the system. Thus, a fast response to changing treatment conditions was possible. Other findings were the absence of drift of the signal at constant water quality and good resistance against mechanical wear of the piston and cell wall of the analyzer.

The response of the SCD to different coagulant doses was evaluated at constant raw water quality and compared to the results of jar tests and zeta potential measurements. The pH was not controlled in order to work under conditions identical to those encountered at the treatment plant. Typical results of these experiments are given in Figure 3.

Within the range of coagulant doses tested, an approximatively linear response of the SCD is measured under these conditions of water quality. Furthermore, a good correlation ($r^2 = 0.94$) between the SCD signal and the zeta potential is observed, a finding that corresponds to theoretical considerations[7] which predict a direct proportionality between streaming current and zeta potential.

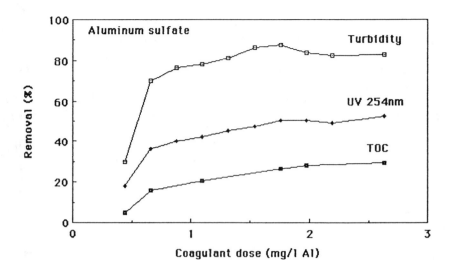
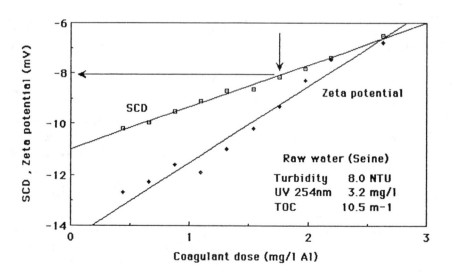

FIGURE 3. Correlation between zeta potential, SCD signal, and jar test results.

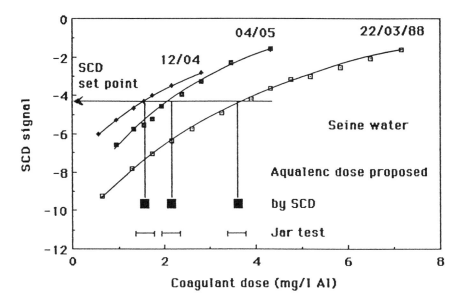

FIGURE 4. SCD response and jar test results for different raw water qualities.

Figure 3a gives the results of a jar test, expressed as removal efficiencies for turbidity and organic matter (UV absorbance at 254 nm and TOC). An optimum turbidity removal is observed at a coagulant dose of about 1.7 mg/L Al. The removal of organic matter increases steadily up to this dose, additional removal requires considerably higher doses. Thus, if turbidity removal linked to a major reduction of organic matter is the treatment goal, a coagulant dose of 1.7 mg/L Al can be chosen.

As shown in the diagram, this optimum dose can be linked to a zeta potential value and also to an SCD signal (about -8, direct reading without units). The latter can be used to control the coagulant dose if charge neutralization is the operative mechanism of particle destabilization.[7] The same SCD signal should always correspond to constant values of zeta potential and thus to the optimum dose of coagulant, and this at different raw water qualities. Figure 4 gives some of our results on Seine river water, where this has been verified for three different water qualities. The comparison of the coagulant dose proposed by the SCD with the jar test data shows a satisfying performance of the analyzer. A constant signal, from now on called "SCD set point," is linked to the dose optima found at different points in time.

Based on these findings, the control of the coagulant injection was automated by using a simple computer algorithm which modifies the coagulant dose stepwise, until the set point is found.

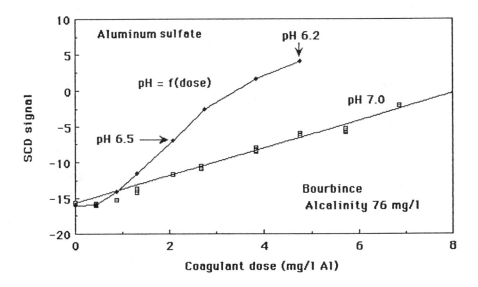

FIGURE 5. SCD response at different pH conditions.

It should be mentioned at this point that the response curves of the SCD are of course not only a function of the coagulant dose, but also of the pH in the coagulated water. Figure 5 gives an example measured with Bourbince river water which represents a lower buffer intensity than the Seine water and where higher doses have been tested. If the pH is held constant, a linear dose-response curve is observed which seems to indicate a direct relation between the charges added to the system and the SCD signal. Without pH control, lower values of this parameter are obtained which result in aluminum species of higher positive charge in the coagulated water. Consequently, the SCD signal is displaced and a nonlinear response curve results which integrates both coagulant dose and pH.

On-Line Control Experiments

Following these first evaluations of the analyzer's performance by batch experiments, it was used for a direct on-line control of the coagulant injection at the pilot Pulsator. The coagulants used were aluminum sulfate and Aqualenc. An initial jar test was used to define the SCD set point corresponding to the optimal treatment dose. The criteria for the latter were optimum turbidity and major organic matter removal.

Data from a typical experimental period where Aqualenc was used as coagulant are given in Figure 6 and show the important variations of turbidity and UV absorbance (254 nm) in the raw water. The curve in Figure 6a gives the coagulant

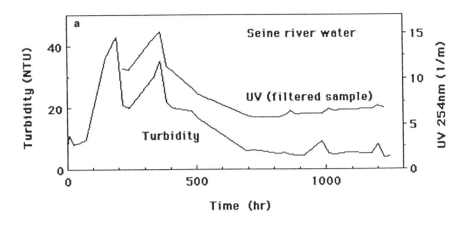

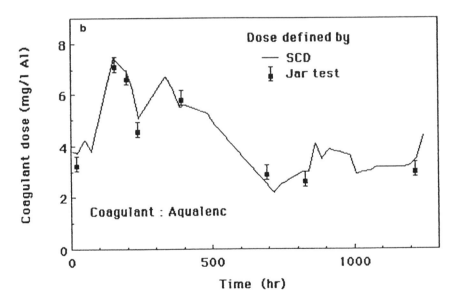

FIGURE 6. Coagulant doses at different raw water qualities, determined by the SCD and jar tests.

dose which has been applied by the SCD. Squares show the optimal coagulant dose determined by jar tests, including a range of uncertainty due to experimental error.

The comparison of the doses applied by the SCD and those determined by jar test show that the analyzer allows a satisfying assessment of their evolution.

Peak demands are identified correctly, as well as fast changes in coagulant dose. No readjustment of the SCD set point was necessary for a period of about 2 months. Similar results were obtained with aluminum sulfate as coagulant during a period of about 3 months.

In order to determine the influence of the continuous dose adjustment by the SCD on the performance of the clarification process, turbidity and UV absorbance (254 nm) were measured in the settled water of the pilot and the industrial size Pulsator (1100 m^3/hr) of the plant. Weekly jar tests were used to adjust the dose of the industrial reactor. The range of parameter values in the raw water, encountered during the experimental period, is shown in Figure 6b.

From the settled water turbidity and UV absorbance data, the cumulative frequencies have been calculated and are shown in Figures 7a and b. It should be noted that the measured data have been normalized using the mean value which corresponds to a cumulative frequency of 50%. This has been done for two reasons. First, the raw water feeding the industrial Pulsator is preozonated (constant dose 1 mg/L O_3), which was not the case for the pilot, in order to avoid additional influences during the evaluation of the SCD. Second, the industrial reactor displayed a slightly better performance at identical operation, due to size effects.

The slope of the lines fitted to the cumulative frequency distributions is representative for the stability of the process. A vertical line would correspond to a constant quality of the settled water; increasing variability of the effluent water quality leads to frequency distributions which deviate more and more from a vertical line.

In what concerns turbidity in the settled water (Figure 7a), the data show an almost identical process stability on the pilot and the industrial reactor. This is probably due to the inherent stability of the Pulsator which tends to buffer water quality variations rather well at least at the water quality changes encountered here. At the same time, the reactor size may be in favor of the industrial process. Under these conditions, on-line control by the SCD does therefore not lead to more stable turbidity values in the settled water as compared to a weekly jar test.

Contrary to these results, the UV data (Figure 7b) show that process stability with respect to this parameter is strongly increased by the on-line SCD control. Whereas the concentration of organic matter in the settled water of the industrial reactor varies considerably, much more stable values are observed on the pilot.

Bourbince River Water

In order to test the SCD on a different surface water, small-scale pilot experiments were run on raw preoxidized water of the Bourbince. Typical raw water data are given in Table 2, showing lower concentrations of alkalinity than

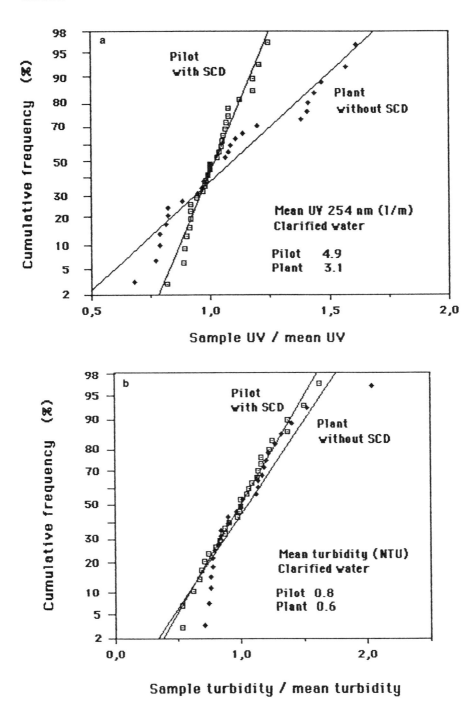

FIGURE 7. Cumulative frequency distribution of turbidity and UV absorbance in the settled water.

Table 2. Raw Water Quality of the Bourbince River (Period 15.3—03.11.89, Seven Samples)

	Minimum	Maximum	Average
Temperature (°C)	11	21	13.8
Turbidity (NTU)	3	21	11.3
Resistivity (ohm cm)	2210	3985	2885
pH at 20°C	7.30	8.3	7.7
Alkalinity (as $CaCO_3$) (mg/L)	76	148	112
Hardness (as $CaCO_3$) (mg/L)	87	156	128
Sulfate (mg/L)	29	66	50.5
Chloride (mg/L)	14	38	25.6
TOC (mg/L)	4.4	6.9	5.43
UV 254 nm (L/m)	9.4	19.3	12.8

the Seine river. At the same time, the TOC content of this water is relatively high.

No direct SCD control of the coagulant dose (aluminum sulfate) was done at this site. Instead, the SCD was installed on the coagulated water of the plant and the response to changing conditions of raw water quality and treatment was measured. By using dose-response curves as they have been shown as an example in Figure 4, the SCD signal was then converted to coagulant doses which were compared to those actually applied. The latter were determined based on jar tests and, especially, the personal experience of the plant operator.

The results from a 3-month period (March 15 to June 21, 1989) are given in Figure 8. Generally, a good correlation between the dose proposed by the SCD and the one actually applied is observed. This is especially true during the period of day 20 to 40, where raw water conditions led to a high coagulant demand. Similarly, a dose increase due to heavy precipitations (day 63) was well-identified. Some deviations between the two approaches of coagulant control are probably due to the fact that the operator did not use a jar test at this moment.

At the same time, however, the evaluation of the data showed that the whole 3-month period could not be modeled by using a single SCD set point. Instead, three different set points (-3.2, $+2.1$, -11.2) had to be used in order to achieve a good agreement with the real dose applied. The duration of the validity of one set point was on the order of 1 month. Response curves of the SCD to increasing coagulant doses, measured at different points in time, confirmed this finding and were used to define the set points given in Figure 8.

Contrary to the Seine river, where constant set points were observed during more than 2 to 3 months, the water quality of the Bourbince seems to lead to more frequent changes in the analyzer's response. Since this requires a closer control, the data available were evaluated with the goal to identify the reason for this behavior. Parameters measured were ion concentrations such as pH, alkalinity, calcium, neutral salts, as well as TOC, turbidity, and temperature. Other parameters, such as the zeta potential, were not analyzed.

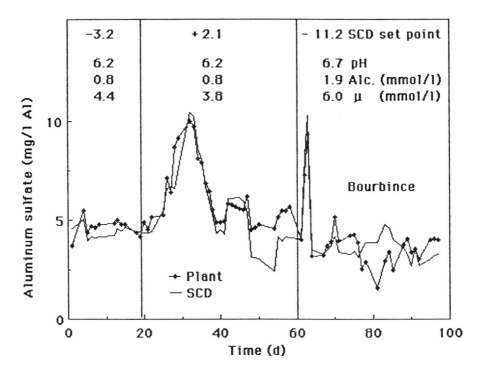

FIGURE 8. Coagulant doses proposed by the SCD and doses actually applied on the plant.

Figure 8 gives the results for the three periods where different set points were observed. The ionic strength (μ) is used to describe the overall concentration of all ions, a coagulant dose of 4.3 mg/L Al is included in the data (pH, SO_4^{2-}). As the results show, no consistent correlation of the set point seems to be present with the parameters pH and alkalinity. However, the set point seems to follow the same trend as the ionic strength. This observation was therefore studied more closely by adding NaCl to the raw water before coagulation during a batch experiment. Results are given in Figure 9.

The data show that rather high values of additional ionic strength are necessary in order to modify the SCD response, since up to about $\mu = 26$ mmol/L the signal remains relatively constant. Only high doses of NaCl lead to a strong modification of the SCD signal, probably due to compression of the electric double layer of the particles. The influence of the general mineralization of the raw water seems therefore not to be responsible here for the changes of the SCD set point. At the same time, literature data gathered using deionized water[7] indicate SCD response modifications at salt concentrations as low as 1 mmol/L. Further research will therefore be necessary in order to confirm these findings and, especially, to identify the reason for the evolution of the set point.

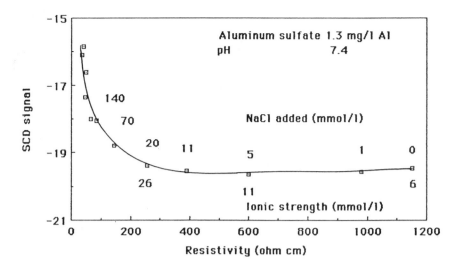

FIGURE 9. SCD response as a function of ionic strength.

CONCLUSION

The experimental evaluation of the SCD, used for the control of the coagulant dose at varying raw water quality, allows the following main conclusions.

The response curve of the SCD to increasing coagulant doses correlates well with zeta potential measurements as expected from theoretical considerations. A constant SCD response is observed for the optimum coagulant dose at different conditions of raw water quality. This signal, which is called SCD set point, can be used for a simple control algorithm, which adjusts the dose until the set point is obtained.

If this approach is used to control the coagulant injection on Seine river water, a good agreement is obtained between the dose applied by the SCD and the optimum dose defined by jar tests. A constant SCD set point is observed for at least 2 to 3 months, that is, no readjustment of the analyzer is required for an extended time period. The SCD can therefore be used without problems to control coagulant injection on this type of surface water.

At the same time, a statistical analysis of the settled water quality shows that more stable concentrations of organic matter (UV 254 nm) are obtained if on-line SCD control is used instead of weekly jar tests. No difference was found for turbidity in the settled water, though, probably due to the inherent stability of the process and reactor size effects.

If the same evaluation was done on another raw water (Bourbince), a good agreement between the coagulant dose proposed by the SCD and the one applied by the plant operator was found as well. However, shorter time periods, where a constant set point can be used, were observed, which were on the order of 1 month. This results in the necessity to control and readjust the SCD more often by using jar tests. If the readjustment becomes more frequent, this may reduce the usefulness of the analyzer. It is, however, not clear actually what leads to the modification of the set point. pH and alkalinity do not seem to be responsible, nor does a change of ionic strength seem to modify the set point unless very high concentrations of salts are added. Further research is therefore necessary in order to clarify this point and to define still better the reliability of the SCD.

ACKNOWLEDGMENTS

We would like to thank Ph. Jacq, chief engineer of the Compagnie des Eaux de Banlieue de Paris, for his cooperation and support of this study.

REFERENCES

1. Brodard, E., Bordet, J. P., Bernazeau, F., Mallevialle, J., and Fiessinger, F. "Modelisation stochastique d'une usine de traitement d'eau potable," paper presented à la 2ème Rencontres Internationales Eau et Technologies avancées, Montpellier, France, 1986.
2. Gerdes, W. F. "A New Instrument — The Streaming Current Detector," 12th National ISA Analysis Instrument Symposium, Houston, TX, May 1966.
3. Bombaugh, K. J., Dark, W. A., and Costello, L. A. "Applications of the Streaming Current Detector to Control Problems," 13th National ISA Analysis Instrument Symposium, Houston, TX, 1967.
4. Smith, C. V., and Somerset, I. J. "Streaming Current Technique for Optimum Coagulant Dose," 15th National ISA Analysis Instrument Symposium, 1969.
5. Galetti, B. J. "Application of the Streaming Current Detector to the Continuous Measurement and Control of Colloidal Systems," 15th National ISA Analysis Instrument Symposium, 1969.
6. Roques, P., and Jersalé, C. "Utilisation d'un détecteur de courant d'écoulement pour asservir le taux de coagulant à la qualité d'eau brute," TSM- L'eau, 80ᵉ année, n° 3, 1985.
7. Dentel, S. K., and Kingery, K. M. "An evaluation of Streaming Current Detectors," *AWWA Res. Found.* (November 1988).
8. Dentel, S. K., and Kingery, K. M. "Using Streaming Current Detectors in Water Treatment," *JAWWA* (March 1989).

CHAPTER 12

Influence of Preozonation on the Clarification Efficiency

Y. Richard

Ozone has been used in drinking water treatment since the beginning of the century in view of the improvement of organoleptic qualities of water, as well as for disinfection. The ozonation ahead a direct filtration was then developed in the 1960s, but the replacement of chlorine by ozone as a preoxidant before a clarification including a sedimentation step appeared at the end of the 1970s. Many experimentations, laboratory tests, and industrial-scale results have been published, and the purpose of this paper is to give some examples of the Lyonnaise des Eaux/Dumez and Degrémont experiences in this field.

METHODOLOGY

The laboratory tests, made with water kept at a constant temperature, included the usual treatability techniques.

Preozonation itself was operated with a concentrated ozone solution (approximately 20 mg L^{-1}) prepared with water from the site studied (same mineral content) after clarification and GAC filtration. This method made it possible to add a small amount of ozone into the raw water studied (for example, 50 mL for 1l, i.e., a treatment rate of 1 mg L^{-1}), while maintaining the same mineral content in the water and without disrupting the coagulation-flocculation process

Table 1. Main Characteristics of the Raw Water of the Houlle

Temperature (°F)	1—25
Turbidity (FTU)	3—20
pH	7.5—8.7
Alkalinity (as $CaCO_3$) (mg L^{-1})	200—300
Cl^- (mg L^{-1})	40—70
SO_4^{2-} (mg L^{-1})	40—100
Ammonium NH_4^+ (mg L^{-1})	0.3—1.5
Oxidability $KMnO_4$ (mg L^{-1}) ($O_2 - H^+$)	12—16
COT (mg L^{-1})	9—14
D.O. (254 nm)	20—35
Fluorescence	4—9
Algae (per ml)	15,000—50,000

to any great extent. The results are expressed per liter of water studied. The quality of the treated water was qualified using the following parameters:

- Turbidity
- Total organic carbon (TOC)
- UV absorbance at 254 nm (UV)
- Humic acid index (HAI)
- Trihalomethane formation potential, measured after chlorination at the rate of 15 mg L^{-1} for 24 hr (measurement with gas chromatography, head space method)
- Eventually, colloidal charge

THE MOULLE PLANT CASE

The Moulle plant is located in the north of France and treats the water of the river Houlle, which is very polluted water. The main characteristics of the water are given in Table 1. This water has an average mineralization, bicarbonate of calcium. It is quite rich in organic matter, the highest levels corresponding to periods of intense developments of plankton. The influence of preozonation on the water quality and the optimum coagulant doses have been tested. The amount of coagulant used for treatment was determined by the jar test technique. After the addition of the coagulant reagent, the zeta potential was measured. After flocculation, the turbidity, the organic matter (TOC), the colloidal charge, the humic acid index, and the potential of haloforms formation were measured. The measurements made it possible to optimize the amounts used in treatment in order to obtain the best possible quality of the treated water.

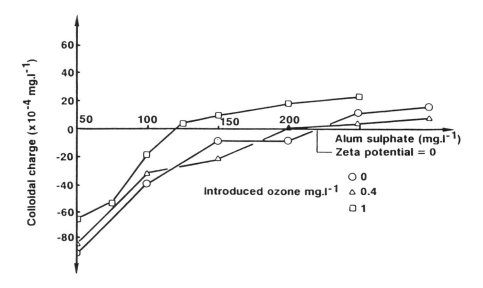

FIGURE 1. Evolution of colloidal charge during jar test after preozonation.

Colloidal Content

Figure 1 shows the evolution of the colloidal titration after preozonation and according to the amount of aluminum sulfate. When no ozone is used, the colloidal content can be cancelled when a dose of 220 mg L^{-1} is utilized, i.e., a dose also corresponding to cancellation of the zeta potential. Preozonation carried out with a dose of 0.4 mg L^{-1} reduces only slightly the necessary dose of coagulant for cancelling the colloidal content, but a dose of 1 mg L^{-1} provides the cancellation for a treatment dose of 120 mg L^{-1} of aluminum sulfate.

TOC

Figure 2 shows that preozonation has no effect on the elimination of total organic carbon. This elimination depends mainly on the aluminum sulfate treatment dose. The TOC is eliminated by adsorption on the metallic hydroxide floc.

Turbidity

Figure 3 indicates the turbidity of the water clarified after preozonation (1 mg L^{-1}), coagulation, and flocculation. It can be seen that the minimum turbidity

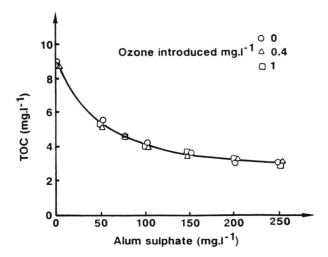

FIGURE 2. Evolution of TOC during jar test.

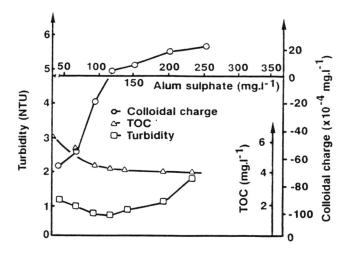

FIGURE 3. Evolution of TOC and turbidity during jar test.

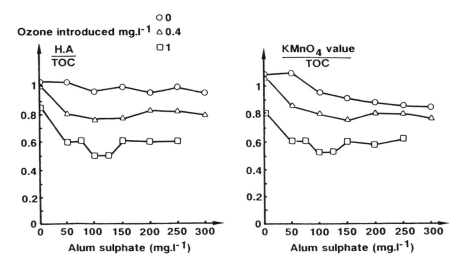

FIGURE 4. Effect of preozonation on organic pollution removal (jar test).

is obtained for an aluminum sulfate treatment dose equaling 125 mg L^{-1}, corresponding to cancellation of the colloidal charge. More than this treatment dose, the turbidity of clarified water deteriorates and the particles remain charged positively.

Organic Matter

The organic matter is measured either by the HAI in the left-hand portion of Figure 4 or by the permanganate value (expressed as mg L^{-1} of oxygen consumed in a hot acid medium). It has just been indicated that ozonation had no effect on the TOC. In order to show up the effect of ozone on oxidability, and particularly on the humic acid index, the evolution of the OM/TOC and HAI/TOC ratios are reported in Figure 4.

As far as oxidability is concerned, the increase of the dose of coagulant without preozonation improves the OM/TOC ratio up to the dose at which colloidal charge is cancelled. The addition of increasing doses of ozone (0.4 and 1 mg L^{-1}) improves the residual value of OM/TOC for smaller doses of coagulant (150 and 120 mg L^{-1}). The HAI varies little when increasing doses of coagulant are added. Ozonation of the raw water considerably improves the HAI/TOC ratio through the use of coagulant doses (120 mg L^{-1}) smaller than the dose needed to cancel the colloidal charge.

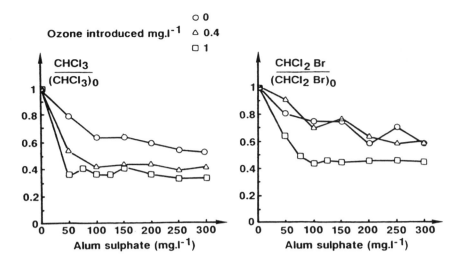

FIGURE 5. Effect of preozonation on the TTHM potential.

Potential of Haloforms Formation

The study of potential of haloforms formation was carried out by the chlorination of the water using high doses of coagulant (15 mg L^{-1}) and a sufficiently long contact time (24 hr) to allow their formation. These conditions are not industrial conditions, but permit a look at the absolute potential of haloforms formation. Figure 5 illustrates these tests. The potentials for the formation of $CHCl_3$ and of $CHCl_2Br$ are expressed by the ratio between the potential obtained after ozonation pretreatment and the potential of the raw water without any ozonation. Coagulation and flocculation considerably reduce the potential of $CHCl_3$ and $CHCl_2Br$ formation: it is not more than 60% of the initial potential for the dose of coagulant needed to cancel the zeta potential (220 mg L^{-1}). Preozonation with a little ozone amount (0.4 mg L^{-1}) improves the elimination of $CHCl_3$ precursors, but the dose must be increased to 1 mg L^{-1} to improve the elimination of the $CHCl_2Br$ precursors.

THE MONT VALÉRIEN PLANT CASE

The Mont Valérien plant is located in the west suburbs of Paris and treats the river Seine water downstream from Paris.

Table 2. Physical-Chemical Characteristics of Seine River Water at Mont Valérien Plant

	During Tests	Changes During Test Year
Resistivity (ohms/cm at 20°C)	2040	1700—2100
Turbidity (NTU)	12	5—80
pH	7.7	7.35—7.8
TOC (mg L^{-1})	4.35	2.5—6
UV absorbance = 254 nm, 4-cm tank	0.435	
Humic acid index (mg L^{-1})	6.2	
Color (mg L^{-1} of Pt.Co)	10	5—25
Cl^- (mg L^{-1})	20	20—25
NO_3^- (mg L^{-1})	17	12—18
NO_2^- (mg L^{-1})	0.3	0.2--0.4
$O\text{-}PO_4^{3-}$ (mg L^{-1})	1.2	0.4—1.6
m-alk (Fr.Deg.)	20.5	20—24
TH (Fr.Deg.)	25.6	25—29
Ca^{2+} (mg L^{-1})	80	
NH_4^+ (mg L^{-1})	0.8	0.4—2
Fe (mg L^{-1})	0.08	0.05—0.25
Mn (mg L^{-1})	<0.01	0.01—0.02

The Raw Water

The main characteristics of the raw water are given in Table 2. Seine river water has average calcium alkalinity with a relatively low turbidity level (2 NTU) and light pollution (organic matter, ammonia). The table also gives the usual range of the various parameters.

Determination of Coagulant Dose

Figure 6 shows the results of the jar test. It can be seen that without preozonation. The optimum aluminum sulfate (expressed as SO_{43} Al_2, $18H_2O$) treatment rate lies between 20 and 30 mg L^{-1}. A treatment rate greater than 30 mg L^{-1} leads to a deterioration in the turbidity of the clarified water (that could be counteracted by adding a flocculation aid), without improving organic matter removal to any marked extent. The treatment rate of 30 mg L^{-1} was therefore adopted for the rest of the tests.

Determination of the O_3 Preozonation Treatment Rate

Figure 7 shows the change in turbidity and TOC of the clarified water for increasing ozone rates in preozonation at a constant coagulant dose of 30 mg L^{-1}.

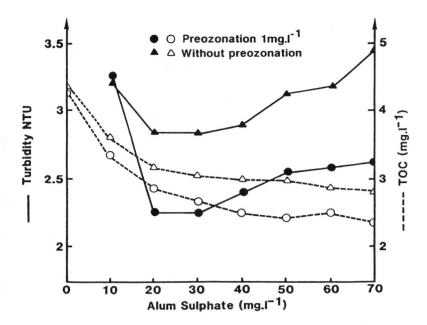

FIGURE 6. Determination of coagulant dose.

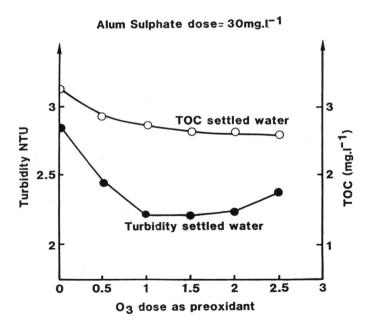

FIGURE 7. Determination of preozonation dose.

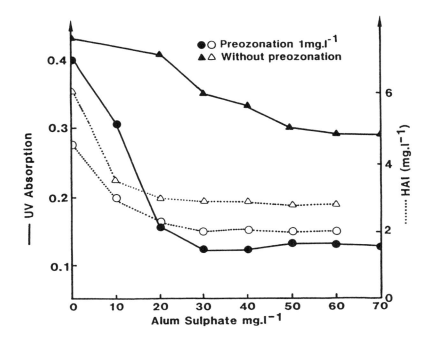

FIGURE 8. Determination of coagulant dose.

The best turbidity level is obtained at an ozone rate of between 1 and 2 mg L^{-1}, but the improvement obtained beyond 1 mg L^{-1} is very small. For economic reasons, it is therefore preferable to adopt a treatment rate of 1 mg L^{-1}. A very slow decrease in the TOC is also noticeable: the organic matter in the water is difficult to oxidize with ozone, or all or at least most has very slow oxidation kinetics. This would imply that the preozonation reactor will have a low yield for a high treatment rate. Increasing the ozone dose to over 1 mg L^{-1} would give a higher ozone loss at the contacting room vent. Figure 6 (determination of the coagulation treatment rate) gives a comparison between the quality of the clarified water with and without preozonation (at a rate of 1 mg L^{-1}) for turbidity and TOC. It shows that while it is difficult to adsorb the organic matter on the aluminum floc before ozonation, there is little improvement after ozonation at the rate of 1 mg L^{-1}. The increase in the aluminum salt treatment rate considerably increases the floc volume and surface area, but only slightly improves the TOC removal. Figure 8 lists the results obtained with clarified water for the UV absorption parameters and humic acid index as a function of the aluminum

Table 3. Combined Effect of Preozonation (1 mg L^{-1}) and Clarification (optimum dose of AS: 30 mg L^{-1})

	Raw Water	Settled Water
TOC	2%	12%
UV absorbance	7%	66%
HAI	25%	33%

sulfate treatment rates. The following conclusions can be drawn from Figures 6 and 7:

- Turbidity reduction — Preozonation makes it possible to reduce the coagulant treatment rate from 30 to 25 mg L^{-1}.
- UV and HAI — The results show that while preozonation provides a marked improvement in the reduction of these two parameters, it is not advisable to reduce the coagulant treatment rate below 30 mg L^{-1} in order to derive the maximum effect from the ozone.
- TOC — Preozonation (at the rate of 1 mg L^{-1}) improves the removal rate of organic matter. However, without O_3, an increase in the coagulant treatment rate above 30 mg L^{-1} does not improve the results; with O_3, it is possible to improve the organic matter removal rate by increasing the aluminum sulfate treatment rate up to 40 mg L^{-1}.

A coagulant treatment rate of 30 mg L^{-1} combined with a preozonation rate of 1 mg L^{-1} would seem to be a good compromise. Table 3 and Figure 9 show the effect of such a combined treatment as far as organic parameters are concerned. They compare the reduction in these parameters with preozonation (1 mg L^{-1}) of the raw water before flocculation and of the clarified water:

- TOC — Preozonation has little effect; it only brings the relative removal rate up from 2 to 12%
- UV adsorption — Relative removal increase from 7 to 66%
- HAI — Relative removal increases from 25 to 33%

In conclusion, it can be seen that there is a certain synergy between the effects of preozonation and coagulation for water at the Mont Valérien plant. However, a compromise must be found between improving turbidity (and so the settling tank's operation) and improving organic matter removal parameters.

Effect on the Total Trihaloform Formation Potential (TTFP)

Figure 10 shows the variation in the trihaloform formation potential as a function of the coagulant treatment rates for a given preozonation rate (1 mg

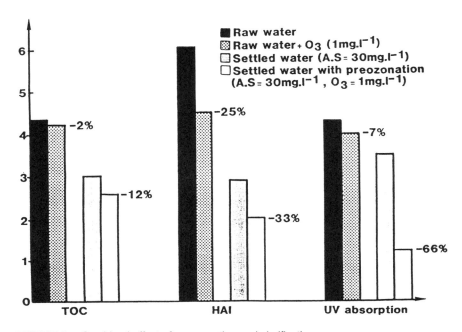

FIGURE 9. Combined effect of preozonation and clarification.

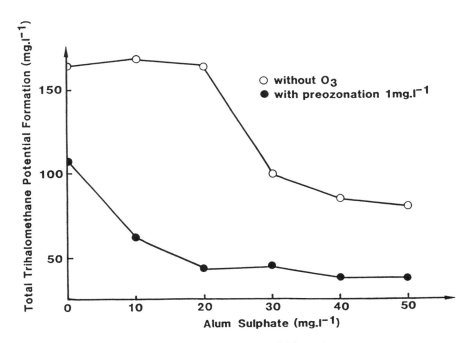

FIGURE 10. Evolution of total trihalomethane potential formation.

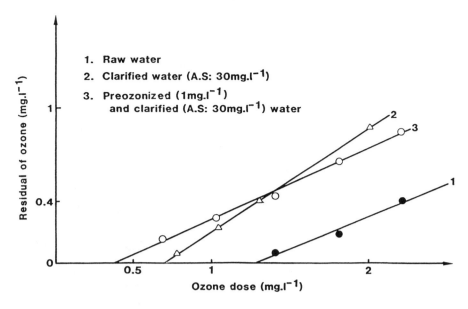

FIGURE 11. Water ozone demand Mont Valérien plant.

L^{-1}). The TTFP falls from 180 to 115 mg L^{-1} for the raw water (that is a 37° reduction). For the water clarified with an aluminum sulfate dose of 30 mg L^{-1}, after ozonation, the TTFP falls from 100 to 40 mg L^{-1}, that is a 60% reduction.

EFFECT OF PREOZONATION ON THE POSTOZONATION OZONE DEMAND

Figure 11 summarizes the results of the ozone demand tests for the raw water and the clarified water with and without preozonation in the case of the Mont Valérien plant. In order to establish the recommended viricidal conditions, it is necessary to inject 2.2 mg L^{-1} of ozone into the raw water. After clarification, the treatment rate is down to 1.3 mg L^{-1}, whether the raw water has been preozonated or not. In that case, the total dose of injected ozone is higher in the case of preozonation (1 + 1.3 = 2.3 mg L^{-1}) than in the case of no preozonation (1.3 mg L^{-1}). The only difference concerns the dose to be injected to obtain a small residual of ozone: 0.4 mg L^{-1} with preozonation and 0.7 mg L^{-1} without preozonation. Figure 12 gives the results obtained on the plant of Vigneux, which treats also the river Seine water, but which is located upstream from Paris. In that case, as the raw water is less polluted, the optimum preozonation dose is only 0.2 mg L^{-1}. The postozone demand in order to establish the recommended

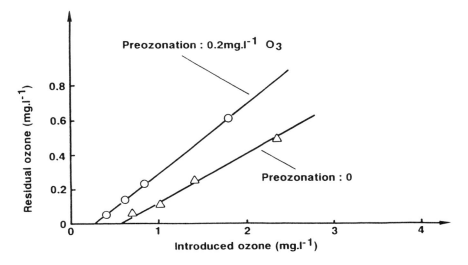

FIGURE 12. Effect of preozonation on treated water ozone demand Vigneux plant.

viricidal conditions is highly reduced, from 2.1 to 1.3 mg L^{-1}. In that case, the total dose of injected ozone is lower in the case of preozonation (1.3 + 0.2 = 1.5 mg L^{-1}) than in the case of no preozonation.

PREOZONATION AND ALGAE REMOVAL

The operating results of the pilot plant of Moulle and of the industrial-scale plant of Mont Valérien showed that the removal of algae with preozonation is similar to the removal of algae with prechlorination. But the preoxidation with ozone does not permit to maintain a residual of oxidant in the settlers: a regrowth of algae can occur, especially in warm and sunny countries. It is then recommended to cover the settler with a simple system, in order to prevent the lighting of the clarification zone.

CONCLUSION

The two case studies which have been presented may lead to summarize the effect of preozonation before clarification:

- Preozonation improves the turbidity and the organic quality of the treated water.
- Preozonation increases the removal of UV absorbance, as well as the total trihalomethane formation potential.
- In some cases, a reduction of the coagulant treatment rate can be observed.
- Preozonation can reduce the total injected ozone dose, but not in all cases.

The differences observed by many authors concerning the effect of preozonation or clarification are probably due to the extremely heterogeneous nature and relative concentration of organic matters in the raw waters. Whenever possible, a laboratory test is recommended to appreciate the real effect of a preozonation treatment step.

CHAPTER 13

Application of Ozone with Activated Carbon for Drinking Water Treatment in China

F. Jinchu and X. Jianhua

INTRODUCTION

Historically, there have been few instances where ozone was used for purification of drinking water in China. Even now, ozone is still not widely used for this purpose. One ozone generator (made in Germany) was used in Xiamen (Amoy) for disinfection of drinking water. Little attention was paid during the period from 1949 to the early 1970s to research on the application of ozone. The reasons are (1) ozonation systems require high capital cost and high energy consumption resulting in high cost of production for drinking water and (2) the history of using chlorine as a disinfectant was accepted and human health hazards caused by trihalomethanes (THMs) were not clearly understood. Only after the 1970s did an increased awareness of the hazards of THMs draw attention to the application of ozonation technology. In 1973, the Shanghai Waterworks Company and Tongji University started to study ozonation of drinking water. In 1979, preozonation was used for the first time for removal of iron and manganese, color, and phenol from water in the Zhoujiadu water treatment plant with a capacity of 30,000 m^3/day. The ozone generator used was a tubular generator made at the treatment plant. The energy consumption for producing ozone was about 20 to 25 kwh/kg. Since the ozone dosage was low, there was inefficient removal of trace organics and ammonia nitrogen.

In the early 1980s, research on drinking water treatment using ozone-biologically activated carbon (ozone-BAC) technology was started. At this time, this technology is being used in eight treatment plants in Beijing, Nanjing, Harbin, Liaoning, and Henan provinces. The total amount of water treated is 600,000 m^3/day. The largest plant is Tiancunshan in Beijing (170,000 m^3/day), and the smallest is 15,000 m^3/day. The total capacity of municipal waterworks in China is 40 million m^3/day, of which only 1.5% used ozone-granular activated carbon (ozone-GAC) technology at the end of 1985.

OZONATION IN CHINESE WATERWORKS

The points of ozone application and dosage vary with the quality of water sources and the treatment objectives. Recently, ozone has been used not just for disinfection, but to remove trace organic and some inorganic compounds. For this reason, the point of application is often in the middle of the treatment process, i.e., after sand filtration and before a GAC filter. Normally, chlorine is applied at the end of the treatment process to maintain a slight chlorine residual in the distribution system. Zhoujiadu is the only waterworks which applies only ozone and does not use ozone-GAC. It has been found that the application of 2 to 3 mg/L of ozone alone is unsatisfactory. Activated carbon following ozonation is needed. In this paper, the use of ozone for drinking water treatment means ozone-GAC or ozone-BAC.

As stated, the point of ozone application and the ozone dosage vary with water quality and the treatment objectives. For example, the Tiancunshan plant in Beijing gets water from Tuanchenghu Lake in the winter, which gets water from Guanting Reservoir. From the reservoir to Tuanchenghu Lake, the water is polluted and the odor may reach class 3 (the worst odor is class 5). The problem for designers of the treatment plant was to remove odor caused by the organic pollutants. A high ozone dosage was required; however, the oxidation by-products of the ozonation process are potentially dangerous if only ozone is used. On the other hand, GAC alone had a very short operation cycle and thus resulting in high operation costs. It was found that a combined ozone-GAC process reduced ozone dosage and increased the GAC operation cycle. The treatment process now in use is

raw water → coagulation → clarification → sand filter → ozonation → GAC filter → chlorination → storage → final chlorination → distribution

Operational parameters for this treatment are ozone dosage, 1 to 1.5 mg/L; filtration through GAC filter, 10 m/hr (the height of the carbon bed is 1.5 m). Under these conditions, odor is reduced from class 3 to class 1 odors. The ozone dosage is increased if odor is above class 4 odors as the capacity of the carbon bed would be severely reduced.

Another example is the Changling Oil Refinery drinking water plant, placed in operation in 1987. This plant uses ozone-BAC to remove phenol, cyanide, oil, sulfide, zinc, and iron. The total ozone dosage is 4 mg/L, divided into two separate steps. The first dose is 2 to 3 mg/L with a 10-min contact time to oxidize most organic compounds, and the second dose is 1 to 2 mg/L after the ozone contactor and before the GAC contactor to further oxidize organic compounds and prolong the BAC operation cycle. The process is as follows:

> Yangtse river water → clarification → sand filtration → pumping station used in industry or from sand filtration → two-step ozonation [(ozone 2 to 3 mg/L) → ozone contactor (ozone 1 to 2 mg/L)] → GAC filter → chlorination → storage

For the waterworks that now use ozone-GAC for the treatment of drinking water, ozone dosages are in the range of 2 to 4 mg/L with 10 to 15 min contact time. In a few cases, contact time exceeds 20 min. Ozone off-gases from the top of the contactor or GAC filter can be recycled to the head of the plant for preozonation. However, as there are often no air pumps with a suitable capacity, the ozone off-gases are passed through a layer of Hopcalite, to decompose the ozone, and the off-gases are vented to the atmosphere. The lifetime of the Hopcalite is about 1 year.

OZONE GENERATORS

All ozone generators that are used were designed and manufactured in China except for those at the Tiancunshan and Changxindian plants, which are from Japan. Tubular generators of the horizontal type were designed by Qinghua University and Shanghai Institute of Chemical Engineering. These ozonators were used for the first time in 1973 for wastewater treatment. Currently, ozone generators are manufactured in factories in Beijing, Shanghai, and Jiangsu provinces. Products manufactured in Shanghai and Jiangsu are series XY, SY, and QH, and are horizontal tubular types. Ozone generators of Yuyuantan Environmental Engineering Facilities Factory (LCF series) are vertical types. Ozone production capacities range from 5 to 1000 g/hr at concentrations of 10 to 20 mg/L. The ozone generator designed by Qinghua University is a vertical plate type and has been manufactured in Shanghai. Air is the raw material for all ozone generators made in China. Energy consumption for ozone production is 20 to 25 kwh/kg; if energy for air cleaning and drying is included, the consumption is 30 to 35 kwh/kg.

First-generation ozone generators were of the tubular type. The high-voltage electrode has a graphite layer coated on the inner wall of a glass tube. This is encased in a stainless-steel tube which is the low-voltage electrode. In this generator, cooling water contacts the stainless-steel tube directly and heat generated by the high-voltage electrode cannot be conducted directly to the stainless-steel tube, but must pass through the glass tube which has poor heat conductivity.

The emission of heat by this type of generator is very poor, which decreases ozone production. Moreover, it requires more energy and also results in the peeling of the graphite layer and puncturing of the glass tube. The first ozone generator installed in the Zhoujiadu plant was of this type. Peeling of the graphite layer and puncturing of the glass occurred frequently. Qinghua University redesigned the generator so that the stainless-steel tube is inside the glass tube, which makes it possible for the glass tube to contact cooling water directly. The heat emission efficiency is increased and puncturing of the glass tube is avoided. Production and energy efficiency improved substantially. Energy consumption is 16 to 17 kwh/kg (Model QH-109). This new model has been produced in batches in Shunyixian Environmental Engineering Facilities Factory since 1980.

THE OZONE-WATER CONTACTOR

The ozone contactor is one of the key factors influencing efficient ozonation. There are several kinds of contactors. Contactors widely used in China are contact tank or contact columns in which ozone flows countercurrently to water. Contact tanks are constructed with reinforced concrete, which costs less, while contact columns are made of more costly stainless steel. The ozone-air mixture from the generator passes through the contactor and microporous diffuser to form microbubbles (less than 1 mm diameter) for thorough distribution throughout the water. The diffuser used in the Zhoujiadu plant is made of microporous carborundum plate, 240 mm in diameter, which can produce microbubbles with diameters in the range of 1 to 2 mm. The ozone utilization is only 75 to 80%, which is not satisfactory.

Now manufacturers are producing a series of microporous titanium plate diffusers. The titanium plates range from 100 to 460 mm in diameter and 2 to 4 mm in thickness. The maximum pore diameter is 250 μm, and porosity is 40 to 50%. The rate of the gas passing through the diffuser is 0.01 L/cm^2-min-Mpa. Information from two operating plants shows that the titanium plate diffuser has the following features: (1) good distribution of microbubbles, (2) fairly fine microbubbles, (3) high efficiency of transfer of gas to liquid phase, and (4) increased ozone utilization rate. Diffusers made of a plastic plate, ceramic plate, or stainless-steel plate do not have the same efficiency as the titanium plates.

Currently, there is active research to increase the ozone utilization rate. For instance, Qinghua University has conducted experiments on a static mixer acting as a contactor. A mixture of gas from an ozone generator and water are whirling in a mixer to form fine bubbles with good mixing and an increased transfer rate. Experiments show that even with a contact time of 1 to 2 sec, ozone utilization can reach 70 to 90%. The advantages of a static mixer are low cost, easy operation, and low maintenance. However, head loss is high, and more experiments are needed to obtain optimal efficiency in large-capacity waterworks.

COST ANALYSIS

Capital costs are dependent on both construction costs and equipment. Costs can vary widely at different locations. For instance, waterworks A is supplying 15,000 m^3/day of water and has a stainless-steel contact column which costs RMB 700,000 (about $190,000), while waterworks B with the same capacity has a concrete contact tank costing RMB 200,000 ($50,000), which is 3.5 times lower than waterworks A. Operation and maintenance costs are mainly dependent upon energy consumption, the activated carbon used, and management practice. A plant in Nanjing using combined ozone-GAC has an operation cost of RMB 220 ($60) per 1000 m^3, but another plant with higher quality raw water has an operation cost of RMB 50 ($14) per 1000 m^3. Thus there is no consistent figure for capital and operation and maintenance costs for waterworks using ozone-GAC.

The capital cost for an ozone generation system, including air cleaning and drying, is about RMB 10,000 ($2700) per 1 kg ozone production per day, but it is only one third to one fourth of that for waterworks not using combined ozone-GAC technology. Energy consumption for producing ozone ranges from 30 to 35 kwh. Assuming that 1 kwh cost RMB 0.11, then the energy cost for production of 1 kg ozone is about RMB 3.85 ($1). A plant with a capacity of 100,000 m^3/day would require an ozone application of 2 mg/L which would require 200 kg of ozone per day. Since additional spare facilities are needed and there are changes in water demand, the total requirement is 300 kg/day. Capital cost for this system size is about RMB 3 million ($810,000). The increased cost is RMB 0.008/m^3 water produced ($0.002) to cover costs for depreciation, maintenance, and labor. If Hopcalite reagent and cooling water system costs are included, the total cost is RMB 0.012 ($0.003). The Tiancunshan plant in Beijing has an increased production cost of RMB 0.005 ($0.014) for GAC, while a plant in Northeast China has an increased cost of RMB 0.015 ($0.004). The difference for these two plants is due to raw water quality and the GAC operation cycle. High cost is the main limitation to increased usage of ozone-GAC technology. Additional research is needed on cost-reduction methods for ozone usage.

RECENT RESEARCH ON OZONATION FOR DRINKING WATER TREATMENT

The research can be divided into three parts:

1. Improvement of equipment used for ozone generation such as generators, air cleaners, air drying facilities, and electrical equipment
2. Increase of ozone utilization rate which includes research on contactors, diffusers, and the combination of ozonation with physical processes, etc.
3. Theory on disinfection and oxidation of organic compounds by ozone

As mentioned before, the ozone generators now used are almost all of a tubular type. The disadvantages are large size, high energy consumption, and low ozone production. For example, LCF series generators, of which Model LCF-100 is the largest capacity, can produce 1000 g ozone each hour, in a volume of 7 m^3, and the weight is 2100 kg. Though there are ozonators of the plate type, they do not have any special advantages. There is a significant lack of research on the improvement of ozone generators. It is reported that a medical institute has designed a small plate ozone generator which has the feature of electric discharge on both sides of the plate. The volume of this generator is only 0.06 m^3, ozone production capacity is 31.5 g/hr, and energy consumption is 17.8 kwh/kg. However, there are no large-scale plate-type products being manufactured. The key point is to enable wide use of ozonation technology for drinking water treatment in China. It is necessary to develop small-size ozone generators, light in weight, consuming less energy and producing ozone at higher rates.

There have been numerous experiments on using ozonation for bacterial disinfection, virus inactivation, oxidation of organic substances (including those which are causing color, taste, and odor), and removal of iron and manganese. Over the years, there has also been research carried out by universities and research institutions on the potential danger from disinfection by-products which may be caused by ozonation for drinking water treatment. The following general conclusions have been confirmed:

1. After ozonation, large molecule organic substances (including mutagens) contained in water are broken down. Some organic substances, such as humic acid, may create new mutagenic substances. So, if ozonation is followed by chlorination for disinfection, THMs may increase since ozonation is unable to decompose THMs. If ozonation is followed by a BAC filter, compounds produced by the ozonation process would be adsorbed or degraded biologically and THM formation after subsequent chlorination would be reduced. Of nine plants using ozonation technology, one (Zhougjiadu) plant is still using only ozonation, but is considering using a combined technology.
2. If ozone is applied at low dosage, no effective removal of ammonia-nitrogen can be expected. If the content of organic nitrogen is high in raw water, ozone will oxidize organic nitrogenous compounds to ammonia-nitrogen compounds. Thus ammonia-nitrogen will be higher in concentration than in the raw water. Systematic research by Harbin Institute of Architectural Engineering and operation in a plant in Nanjing support this conclusion. In this plant, ammonia-nitrogen is 10% higher after ozonation than in raw water. In the dry season, when raw water is very poor, ammonia-nitrogen concentration in the water treated with ozone may be several times higher than in the raw water.

In addition to the application of combined ozone-GAC or ozone-BAC technology, research on combined ozonation-ultrasonic and ozonation-ultraviolet technology are also being studied.

Xie Qingyuan conducted experiments on disinfection by ozonation-ultrasonic technology (the frequency of the ultrasonic wave is 20 kHz). The experiment showed that this process can accelerate *E. coli* destruction and 50% ozone can be saved by the process. The mechanism involved in the disinfection process is a problem that needs to be studied.

Lü Xiwu has conducted experiments on the application of ozonation-ultraviolet technology to remove trace organic compounds from water. His experiment showed that this combined technology, with the photochemical oxidation, can effectively remove chloroform, benzene, toluene, ethyl benzene, hexachlorobenzene, pentachlorophenol, and others. The results from analysis using ultraviolet scanning and capillary gas chromatographic techniques showed that removal efficiency for a broad spectrum of organic substances was 50%; the result from a mutagenic test on SOS was negative. It is expected that for advanced treatment of small amounts of drinking water, that ozonation can be useful. The high cost of ozone prevents this technology from being applied in large-scale water treatment in the near future.

In conclusion, it is expected that the application of ozonation and various combinations of ozonation with physical or biological processes to remove trace organic substances from drinking water will increase.

CHAPTER 14

Basic Concepts for the Choice and Design of Ozone Contactors

M. Roustan, E. Brodard, J. P. Duguet, and J. Mallevialle

ABSTRACT

The purpose of this paper is to present the basic concepts for the choice and design of ozone contactors. Based on fundamental equations concerning gas-liquid mass transfer, reaction, and hydrodynamic behavior, a method for the design of ozone contactors is proposed. By using these concepts a new ozone contactor has been developed.

The importance of hydrodynamic behavior is clearly shown for the CT concept used for the inactivation of virus or cysts.

INTRODUCTION

In the field of water treatment, ozone is used for various purposes: oxidation of organic or mineral compounds, disinfection, pretreatment before coagulation, or filtration. The ozonation process is fairly complex and many factors affect the ozone mass transfer. Unfortunately, few data are available for calculating ozone contactor, so the design criteria are based on empirical methods. In several early published papers,[1-5] the fundamental concepts of gas-liquid mass transfer, hydrodynamic behavior, and chemical reaction have been introduced in order to give some elements for a better design of ozone contactors.

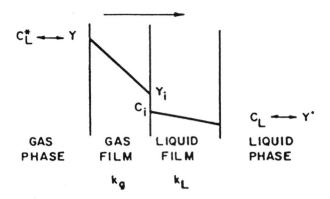

FIGURE 1. The Lewis-Whitman double film theory.

Since parameters such as bubble diameter, interfacial area, etc., are difficult to determine, an overall approach is proposed, i.e., the ozone mass balance on full-scale contactors in order to determine overall parameters: volumetric mass transfer coefficient $k_L a$, ozone demand, transfer efficiency. This approach, with associated equations, has permitted development of a procedure for the design of industrial contactors and for the study of sensitivity of one parameter on ozone transfer.

Bubble diffuser contactor is presently the predominant technology. Nevertheless other technologies may be used. A new process, the Deep U Tube, has been developed and a comparative study is presented concerning the CT concept.

FUNDAMENTALS OF OZONE MASS TRANSFER

Physical Absorption

Ozone transfer will be treated by using the well-known Lewis-Whitman double film theory,[1-3] which explains the rate of transfer of a material from one phase to another. Figure 1 describes the situation.

The mass transfer equation can be written:

(amount transferred per unit time)

= (transfer coefficient) (exchange surface) (exchange potential) (1)

The ozone is slightly soluble in the water, so all the resistance occurs in the liquid phase. The equation becomes:

$$N = k_L S (C_L^* - C_L) = k_L a (C_L^* - C_L) V$$

$$\text{kg/sec m/sec m}^2\text{ kg/m}^3\text{ kg/m}^3\text{ m/sec m}^2/\text{m}^3\text{ kg/m}^3\text{ kg/m}^3\text{ m}^3 \qquad (2)$$

where C_L is the ozone concentration in water
C_L^* is the equilibrium ozone concentration (saturation) with a gas phase containing an ozone concentration C_g

The equilibrium law is

$$C_L^* \,(\text{kg/m}^3) = mC_g (\text{kg/m}^3)$$

m is a function of temperature and pressure (for example at 20°C m = 0.35 for 1 atm and 0.70 for 2 atm).

In a gas-liquid exchanger, it is often difficult to determine the exchange area S. So it is better to express S per unit volume of liquid V, i.e., a = S/V, the interfacial volumic area (m²/m³). The coefficient $k_L a$ (s^{-1}) is the main characteristic of a gas-liquid exchanger.

Absorption with Chemical Reaction

When ozone reacts with some organic compounds in water according to a kinetic law, it is important to determine whether this reaction increases the amount of ozone transferred. This increasing is characterized by the enhancement factor E which is the ratio between the absorption flux in presence of a reaction and the physical absorption flux with $C_L = 0$. Equation 2 becomes:

$$N = E k_L a C_L^* V$$

Two criteria serve for the determination of the factor E:

- Hatta number defined by
  $Ha^2 = k\, Do_3/k_L^2$
- R number, defined by
  $R = k\beta/k_L a$

for a first-order reaction
β liquid holdup
Do_3 diffusivity of ozone into the liquid
$k_L a$ volumetric mass transfer coefficient

Hatta number indicates where the reaction takes place, either into the bulk of the liquid or into the liquid film. R number represents the ratio of ozone consumed into the liquid by chemical reaction and the transferred quantity:

- Very slow reaction — Ha < 0.3 R < 0.1
  The reaction takes place into the bulk of the liquid and the appropriated reactor is a reactor with great liquid holdup β (bubble column, Deep U Tube).
- Slow reaction — Ha < 0.3 R > 0.1
  A part of the absorbed gas reacts before leaving the reactor. A particular case occurs when R > 10, then the concentration C_L is near 0 but E = 1 (bubble column, agitated vessel, Deep U Tube)
- Fast reaction — Ha > 3
  A very important fraction of ozone reacts into the diffusional liquid film and the concentration in the bulk of the liquid C_L is 0. In this case E > 1 and the most important is the interfacial area created (the liquid holdup β has no influence), so the best reactor is the plate or packing column.

HYDRODYNAMIC BEHAVIOR

The two ideal behaviors for a phase (gas or liquid) are plug flow or perfectly mixed. The real behavior of a contactor is generally situated between these two ideal behaviors. It is possible to modelize the behavior by using different models.[6] For that it is necessary to perform tracer test in a full-scale unit. Two common methods are generally employed:

- The "pulse" method, which consists of adding to the water, at the entrance, an instantaneous dose of tracer. The response curve at the outlet gives the "C" or "E" curve (external age distribution):

$$E = \frac{j^j}{(j-1)!} \theta^{j-1} e^{-j\theta}$$

$$\theta = t/\tau$$

- The "step" method, which consists to add to the water, at the entrance, a continued quantity of tracer (concentration C_o). The respond curve at the outlet C/C_o vs time gives the "F" curve:

$$F = 1 - e^{-j\theta} \sum_{i=1}^{} (j\theta)^{i-1}/(i-1)!$$

A relationship exists between the two respond curves:

$$F = \int E dt \quad E = dF/dt$$

Table 1. Comparison of Different Flow Models

Order	Perfectly Mixed	Plug Flow
0	$X = \dfrac{k_0}{C_i}$	$X = \dfrac{k_0}{C_i}$
1	$X = \dfrac{k_1 \tau}{1 + k_1 \tau}$	$X = 1 - e^{-k_1 \tau}$
2	$\dfrac{X}{(1 - X)^2} = K_2 \tau\, C_i$	$\dfrac{X}{(1 - X)^2} = k_2 \tau\, C_i$

Note: τ mean residence time, $X = (C_i - C_o)/C_i$, C_i inlet, and C_o outlet.

It is very important to know the hydrodynamics of the phases for two reasons:

1. Table 1 gives the values of the fractional conversion X of a given reactant for the two ideal behaviors and for three orders of reaction.
   For a first-order reaction, with a plug flow, X is greater than that obtained with a well-mixed flow, for the same mean residence time.

   if $k\tau = 9$ then X plug flow   = 0.99

   X perfectly mixed = 0.90

   With a value of X fixed (X = 0.9) $k\tau$ plug flow = 2.3 and $k\tau$ well-mixed flow = 9. So the mean residence time is smaller than this obtained with a well-mixed flow.

2. Recently the U.S. EPA proposed a guide for disinfection requirements "for public water systems using surface water sources". The concept of CT has been introduced in which C is the ozone concentration in the liquid phase and T the detention time. Several CT values for ozone inactivation of different products have been proposed. For example at 15°C:

   Giardia Cysts pH 6-9, 3 Log removal CT = 1.3

   Viruses pH 6-9, removal greater than 4 Log CT = 0.4

   (C g/m^3 T min).

The detention time T used for the determination of CT value is the time corresponding to when the "F" curve is set equal to 0.10.

The influence of flow pattern models on the relationship between T and mean hydraulic residence time τ is shown here:

Complete stirred tank reactor	$T = 0.10\, \tau$
Three perfectly mixed reactors in series	$T = 0.37\, \tau$
Six perfectly mixed reactors in series	$T = 0.52\, \tau$
Plug flow reactor	$T = \tau$

So for the same hydraulic residence time τ, the reactor which has a behavior nearly plug flow, is more advantageous.

Table 2. Main Characteristics of Contactors

	Volume (m³)	Height (m)	Liquid Flow Rate (m³/hr)	Gas Flow Rate (m³/hr)	Cge (g/m³)
Aubergenville	1460	5.5	2500—5500	137—300	9.3—15.2
Croissy	319	4.6	630—1380	140	6.4—10.5
Moulin les Metz	245	4.3	1150—2200	140—320	7—14.6
Vernouillet	70	5	260—377	39	9.6—14
Villeneuve la Garenne	410	4.6	680—1380	87—174	8.5—15.2

Table 3. Types of Water

Aubergenville	Nitrified, decarbonated, filtered, underground water
Croissy	Nitrified, underground water
Moulin les Metz	Chlorinated, coagulated, filtered surface water
Vernouillet	Nitrified, underground water
Villeneuve la Garenne	Nitrified, underground water

MASS BALANCE OF OZONE IN DIFFERENT OZONE CONTACTORS

Mass balance of ozone has been made in several industrial contactors in order to determine ozone utilization in the contactor: bubbles contactors and Deep U Tube.

Bubbles Contactors

Each contactor is made of two chambers. Ozonated air is introduced into the chambers by porous discs. The main geometric characteristics of the contactors and water quality are given in Tables 2 and 3.

The schematic of the contactor and the associated symbols used in the equations are shown in Figure 2.

The total quantity of ozone transferred OT (g hr^{-1}) from air to water per unit of time is determined with mass balance on the gas phase: OT = G (Cge − Cgs). For each chamber we can write the following equations where:

$$Q_1 = OT_1/L = G_1/L \, (Cge - Cgs_1) = (Cl_1 - Cl_0) + X_1$$

$$Q_2 = OT_2/L = G_2/L \, (Cge - Cgs_2) = (Cl_2 - Cl_1) + X_2$$

Equations may be also written:

$$Q_1 = k_L a)_1 \cdot \Delta C_{mL})_1 V_1/L$$

$$Q_2 = k_L a)_2 \cdot \Delta C_m)_2 V_2/L$$

where $k_L a$ is the volumetric mass transfer coefficient
ΔC_{mL} is the mean exchange potential on the reactor

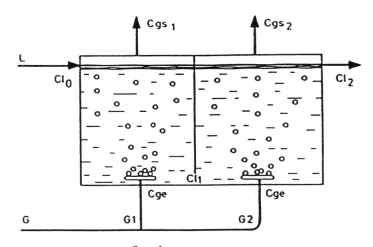

L	: liquid flowrate ($m^3.h^{-1}$)
G	: gas flowrate ($Nm^3.h^{-1}$)
G1	: 1st chamber gas flowrate ($Nm^3.h^{-1}$)
G2	: 2nd chamber gas flowrate ($Nm^3.h^{-1}$)
Cge	: inlet ozone concentration in the gas ($g.Nm^{-3}$)
Cgs1	: oulet ozone concentration from the 1st chamber ($g.Nm^{-3}$)
Cgs2	: oulet ozone concentration from the 2nd chamber ($g.Nm^{-3}$)
Cl_0	: ozone residual in the water before the first chamber ($g.m^{-3}$)
Cl_1	: ozone residual in the 1st chamber outlet ($g.m^{-3}$)
Cl_2	: ozone residual in the 2nd chamber outlet ($g.m^{-3}$)

FIGURE 2. Scheme of ozone contactor with the main symbols.

These mass balances made on the full-scale contactors have been permitted to determine:

- The ozone demand X, by the different waters
- The volumetric mass transfer coefficient $k_L a$
- The ozone transfer efficiency R = 100 (Cge − Cgs)/Cge

	X g/m³	$k_L a$ mn⁻¹	R%
Aubergenville	0.25—0.50	0.003—0.03	86—91
Croissy	0.50—1	0.03	74—81
Moulin les Metz	0.8—1.2	0.02—0.07	68—85
Vernouillet	0.7—0.8	0.05—0.08	85—86
Villeneuve la Garenne	0.6—1.2	0.009—0.02	60—73

In Figure 3 we have reported the values of k_La vs gas flow rate expressed per unit of cross area of the contactor (44 experimental data). The correlation is expressed as:

$$k_La = 0.013 \; U_G^{0.95} \quad \text{at} \quad 12°C$$

$$(mn^{-1}) \qquad (m/hr)$$

Very little mass data are currently available in the literature. So the comparison is difficult. These data may be compared to the values which would be obtained from the different correlations concerning the gas/liquid mass transfer, which are function of bubble diameter, film mass transfer coefficient, gas holdup, interfacial area, etc. Several models have been proposed.[1,2,4,5]

The most important result of the application of ozone mass balance is to define a procedure[7,8] for the design of a similar type of ozone contactor equipped with an identical gas diffusion system (porous plates) (see Figure 4). With this approach, it is possible to study the sensitivity of several factors affecting ozone transfer, by using computer and iterative methods.[8]

Example: Influence of the height H of the contactor, when volume and liquid flow rate are maintained constant:

	X g/m³	G m³/hr	R%	A g/m³	$C_L = 0.2$ g/m³
	0	23	77	0.26	Cge = 13 g/m³
H = 4.3 m	0.5	83	74.3	0.94	V = 116 m³
	1	145	73.3	1.64	L = 1150 m³/hr
	0	20	87.6	0.23	
H = 6 m	0.5	72	85.9	0.82	$A = \dfrac{Cge \cdot G}{L}$
	1	124	85.2	1.41	

DEEP U TUBE

Recently, by using the concepts of mass transfer and hydrodynamic, a new ozone contactor, the Deep U Tube (DUT), has been developed.[9,10] Figure 5 gives a typical schematic diagram of a full-scale plant. It is composed of a vertical tube in a cylindrical closed tube. The vertical tube receives inlet water and gas at the top. The high turbulence in the inner tube breaks the gas into bubbles. As the liquid and bubbles go downwards, the pressure increases and also the equilibrium saturation concentration C_L^*. At the bottom the water and the bubbles flow upwards in the annular tube.

A theoretical approach has been made concerning the hydrodynamic behavior of the downward gas-liquid flow (gas and liquid holdups, pressure drop in bubbly flow).[10] So, the hydrodynamic parameters may be predicted.

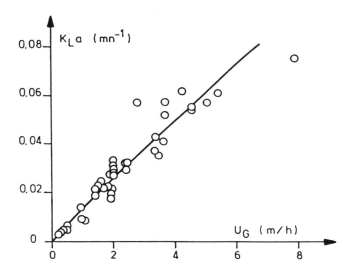

FIGURE 3. Evolution of $k_L a$ vs gas flow rate U_G at 12°C.

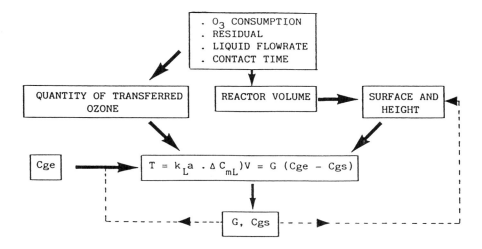

FIGURE 4. Procedure for calculating the optimal design and operating conditions of an ozone contactor.

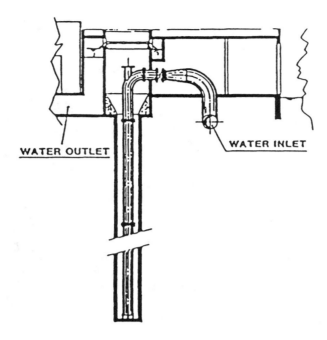

FIGURE 5. Industrial Deep U Tube.

A mass balance of ozone has been made in a full-scale plant (Le Pecq, nitrified underground water). The geometrical dimensions are as follows:

External diameter	1.2 m
Inner tube diameter	0.55 m
Total height	20 m
Total volume	40 m^3
Liquid flow rate	1200—1500 m^3/hr
Gas flow rate	65—160 m^3/hr
O_3 concentration in gas phase	7.5—18.7 g/m^3

The results are presented in Table 4.

From a point of view of ozone mass transfer efficiency, the DUT has R values greater than the bubble contactor, due essentially to the height and/or pressure effects.

CT CONCEPT AND HYDRODYNAMIC BEHAVIOR

The bubble diffusor ozone contactors constituted by two chambers in series can be modelized by three reactors in series so the detention time $T = 0.37\ \tau$.

Table 4.

L (m³/hr)	G (m³/hr)	Cge (g/m³)	Cgs (g/m³)	Cl (g/m³)	R%	O₃ consumed (g/m³)
1200	80	7.5	0.1	0.2	98.6	0.3
	80	15	0.2	0.6	98.6	0.4
1500	65	12.7	0.2	0.36	98.4	0.18
	80	8.7	0.1	0.2	98.7	0.26
	80	18.7	0.25	0.65	98.7	0.34
	120	18.7	0.30	1.23	98.4	0.26
	160	17.9	0.30	1.39	98.3	0.49

For the Deep U Tube (Le Pecq) tracer test permit to obtain $T = 0.55\,\tau$. With the Deep U Tube without cylindrical uppersection $T = 0.90\,\tau$.
So, for a $CT = 0.40$:

$$\text{Bubble contactor} \quad 0.37\, C\tau = 0.4 \to C\tau = 1.1$$
$$\text{DUT} \quad 0.9\, C\tau = 0.4 \to C\tau = 0.44$$

- If ozone concentration C is the same for the two reactors, the mean residence time (V/L) in a bubble contactor is greater than in a DUT (2.5 times), which is important for the capital costs of a full-scale unit.
- If mean residence time τ is the same for the two reactors ($\tau = 4$ min), the ozone residual concentration in a bubble reactor ($C = 0.27$ g/m³) is greater than in a DUT ($C = 0.11$ g/m³), which is important for the operating costs for ozone (applied ozone dosage is less for a DUT).

CONCLUSION

The introduction of mass transfer, reaction, and hydrodynamic behavior has permitted to give some elements for a better design of ozone contactors.

Ozone mass balances have been carried out in several industrial bubble diffuser contactors. A correlation between the overall volumetric mass transfer ($k_L a$) and the superficial gas velocity (U_G) has been obtained and a procedure has been developed in order to design and optimize these contactors. The experimental data obtained for a new ozone contactor (Deep U Tube) have shown that transfer efficiency may be very important. A comparative study of the two technologies has shown the importance of the hydrodynamic behavior on the removal efficiency of different organisms (virus and cysts). A plug flow reactor is more advantageous for this problem.

REFERENCES

1. A. G. Hill, and H. T. Spencer. "Mass Transfer in a Gas Sparged Ozone Reactor," in *Proceedings of the First International Symposium on Ozone for Water and Wastewater Treatment*, R. G. Rice and M. E. Browning, Eds. (1973), pp. 367-381.
2. J. Mallevialle, M. Roustan, and H. Roques. "Détermination Expérimentale des Coefficients de Transfert de l'Ozone dans l'Eau," *Trib. CEBEDEAU*, (1975), p. 377.
3. W. Masschelein, Ed. In: *Ozone Manual for Water and Wastewater Treatment* (New York: John Wiley & Sons, 1982).
4. I. Stankovic. "Comparison of Ozone and Oxygen Mass Transfer in a Laboratory and Pilot Plant Operation," *Ozone Sci. Eng.* 10:321-338 (1988).
5. K. L. Rakness, R. C. Renner, B. A. Hegg, and A. G. Hill. "Practical Design Model for Calculating Bubble Diffuser Contactor Ozone Transfer Efficiency," *Ozone Sci. Eng.* 10:173-214 (1988).
6. O. Levenspiel. In: *Chemical Reaction Engineering*, 2nd ed. (New York: John Wiley & Sons, 1972).
7. M. Roustan, J. P. Duguet, B. Brette, E. Brodard, and J. Mallevialle. "Mass Balance Analysis of Ozone in Conventional Bubble Contactors," *Ozone Sci. Eng.* 9:289-297 (1987).
8. M. Roustan, J. P. Duguet, E. Brodard, and J. Mallevialle. "Ozone Mass Balance in Different Industrial Ozone Reactors: Bubble Column and Deep U Tube," in *Proceedings 9th Ozone World Congress*, Vol. 2, New York, NY, June 1989, pp. 543-550.
9. E. Brodard, J. P. Duguet, J. Mallevialle, and M. Roustan. "A New Method to Dissolve Ozone into Water: Deep U Tube," *Environ. Tech. Lett.* 7:469-478 (1986).
10. M. Roustan, K. Touhami, A. Liné, E. Brodard, J. P. Duguet, and J. Mallevialle. "Theoritical Approach and Experimental Results Obtained for a New Ozonation Gas-Liquid Reactor — the Deep U Tube," Ozone in Water Quality Management, Enprotech./IOA Conference, London, October 18 to 20, 1988.

CHAPTER 15

Organic Removal by Ozonation and GAC Adsorption

Y.-B. Feng, J.-S. Feng, and Y.-N. Wong

INTRODUCTION

Ozonation and granular activated carbon (GAC) filtration as polishing treatments are used around the world in many drinking water treatment plants for the removal of tastes and odors, micropollutants, etc. In order to improve the quality of the Macau drinking water, a new water plant has been proposed. An ozone/GAC pilot plant is being run to optimize unit operations for the new treatment plant. Water quality parameters such as total organic carbon (TOC), optical density $(OD)_{254}$, trihalomethanes (THMs), THM formation potential (THMFP), closed-loop stripping analysis (CLSA)-GC/MS, and chemical oxygen demand (COD) are measured frequently to develop an understanding of how to optimize the water treatment plant. Complementary ozonation tests performed in a semibatch reactor provide kinetic data which are useful for the design by the Lyonnaise des Eaux/Dumez Expert System.[1]

The results of this research program will provide information for further improvement in the quality of Macau drinking water and offer technical support for the eventual adoption of ozonation and GAC filtration techniques in all of Macau's water treatment plants and other areas of southeast Asia.

EXPERIMENTAL EQUIPMENT AND METHODS

Pilot Plant

The pilot-plant scheme is shown in Figure 1. It includes two ozone contacting columns, one GAC filtration column, one sand filtration column, and an ozonator. The height of the ozone column is 6 m. The height of the GAC and sand columns are both 2 m. The water source to the plant is sand-filtered water delivered by the industrial treatment plant, as shown in Figure 2. Operational conditions of the pilot plant are:

1. Ozone column: flow rate of water 1450 L/hr, flow rate of air 2 × 73 normal L/hr corresponding to a total contact time of 8 min; the air/water flow rate ratio is 0.11.
2. Filtration column: bed depth of carbon or sand bed of 1.0 m; flow rate per filter 400 L/hr corresponding to a filtration velocity of 8 m/hr
3. Concentration of ozone in air 15 mg/L.

Methods

Determination of the ozone concentration in air and water: ozone concentrations are measured at the inlet and outlet of the ozone contacting columns. The difference between the two values is the amount consumed or utilized by reactions in the aqueous medium. The iodometric method[2,3] is used for the determination of ozone in the gas phase. The indigo method[4,5,6] is used for the determination of dissolved ozone concentration in the water.

Optical Density Measurement

In the case of ozonated water samples, residual ozone is stripped by nitrogen. A UV-visible spectrophotometer is used to determine the optical density at 254 nm.

Closed-Loop Stripping Analysis (CLSA) Method[8]

Nonpolar hydrophobic substances are stripped from water samples, trapped by a small carbon trap in a hermetically closed system. Organic compounds are eluted by carbon disulfide from the carbon and analyzed by capillary gas chromatography and mass spectrometry.

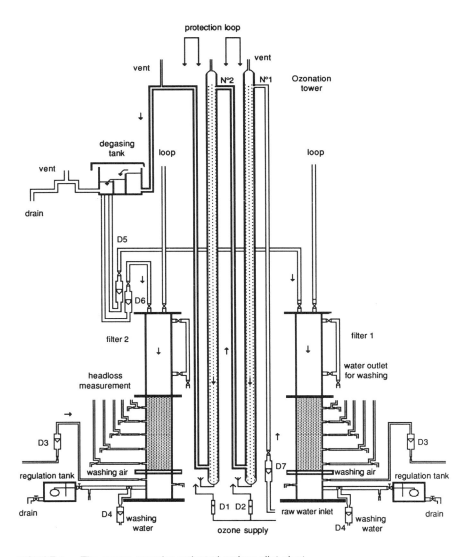

FIGURE 1. The ozone-granular activated carbon pilot plant.

Flavor Profile Analysis Method

Six people were selected among the laboratory staff as panelists for taste and odor analysis. The seven-point intensity scale[1-12] in drinking water analysis was used, where 1 (threshold), 2,4 (slight), 6,8 (moderate), 10, and 12 (strong).

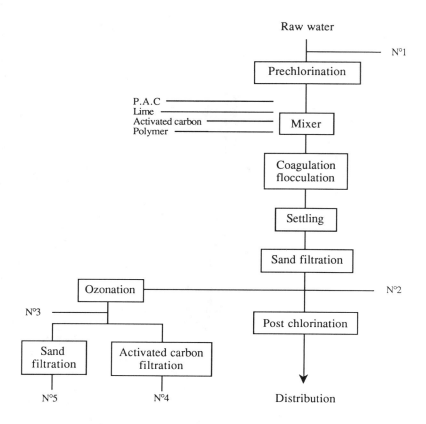

FIGURE 2. General scheme of the 90,000 m³/day industrial and pilot plants.

THM Analysis Methods[10]

The water sample is collected in a 50-mL brown glass bottle. When the bottle of water sample is heated, the THM in the water sample will vaporize into headspace. The THM-containing gas in the headspace is injected into the gas chromatograph for analysis.

The COD and NH_4^+, NO_3^-, NO_2^-, etc. concentrations were determined using the methods described in "Standard Methods for the Examination of Water and Wastewater".[10]

RESULTS AND DISCUSSION

The Kinetic Constants and "Stoichiometric Coefficient"

Natural water is composed of very complex organic matter. In order to characterize the reactivity of ozone with the organic content, it is possible to determine kinetic constant related to a surrogate parameter — the optical density at 254 nm.[11] The decrease in optical density (OD) at 254 nm during ozonation vs time can be modeled by two lines on a semilogarithmic scale. These two lines represent first-order kinetics. Their slopes are, respectively, equal to K_1 and K_2. An ozone expert system developed by the Center of International Research for Water and Environment (Lyonnaise des Eaux/Dumez)[1] permits the design and optimization of the operating conditions of the ozonation process based upon these constants. In order to run the models, kinetic data and the "stoichiometric coefficient" (OD_{254} removed/ozone consumption) in relation to a specific water are needed. The operating conditions for kinetic constants and stoichiometric coefficient are

- Semibatch contactor, 10-L capacity
- Volume of ozonated water, 6 L
- Initial concentration of ozone in gas, 13.7 mg/L
- Water temperature, 26°C

Figure 3 shows an example of the semilogarithmic evaluation of OD_{254} nm. The kinetic constants are determined by the linear regression method:

K_1: y = -0.097x + 0.271 K_1 = -0.097 min^{-1}, R = 0.927

K_2: y = -0.022x $-$ 0.367 K_2 = -0.022 min^{-1}, R = 0.937

For a 50% OD_{254} removal, the value of the stoichiometric coefficient is 4.2 mg O_3/L per unit of removed OD_{254} (expressed per meter of cell length). This data will be used for the design of the ozone contactor which may be used in Macau. The results of this design will be presented in future publications.

Pilot-Plant Results

Ultraviolet absorption at 254 nm (OD_{254}): the average value of OD_{254} of the sand-filtered water is 3 (Figure 4). Ozone reacts with organic matter such as polyhydroxyaromatic compounds by cleaving the rings resulting in OD_{254} removal. Figure 4 shows that ozonation of the sand-filtered water results in a 40 to 60% OD_{254} removal. After the GAC filtration, the OD_{254} removal reaches more than 80%. The OD_{254} values of GAC-filtered water increase slowly as a function of the time of application.

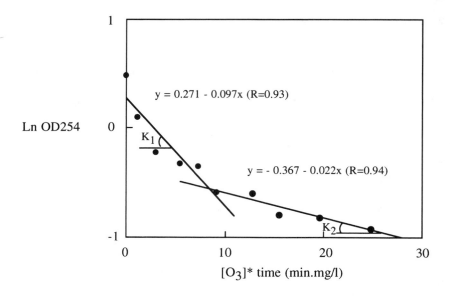

FIGURE 3. Semilogarithmic evolution of optical density at 254 nm of the Macau clarified water as a function of the ozonation time.

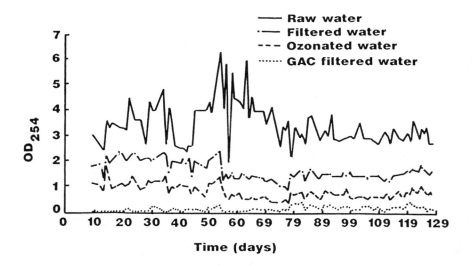

FIGURE 4. Evolution of the optical density at 254 nm along the industrial and pilot-plant steps.

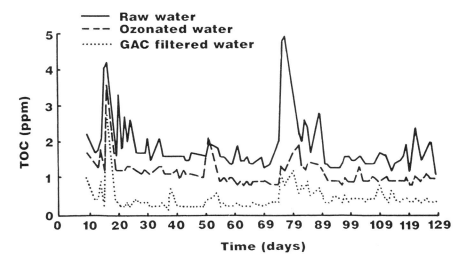

FIGURE 5. Evolution of TOC along the industrial and pilot-plant steps.

Total Organic Carbon

After clarification, the organic content of the water is low and the average TOC value is near 1.5 mg/L (see Figure 5). Obviously, ozonation does not change the TOC value, since ozone only degrades high molecular weight compounds to polar compounds such as aldehydes, ketones, and carboxylic acids. Ozone does not mineralize organics to CO_2 with the ozone doses used for water treatment. The TOC values decrease dramatically after the O_3/GAC treatment. In particular, the removal efficiency reaches 80% during the first 60 to 90 days, after which the removal remains constant.

Chemical Oxygen Demand

COD is slightly reduced during ozonation and a large reduction occurs after ozonation and GAC filtration. Initially, COD values of the GAC effluent were stable and tended to increase progressively with time.

Trihalomethanes and Trihalomethane Formation Potential (THMs and THMFP)

Table 1 shows that the THM concentration is very low in raw water and increases after sand filtration. This is a result of prechlorination. As expected,

Table 1. Evolution of THM Values Along the Treatment Line

Sample THM Date (ppb)	Raw Water	Filtered Water	Ozonated Water	GAC-Filtered Water
April 10	2.2	2.7	2.7	0.6
May 19	0	12.6	12.1	0
May 26	0	5.0	4.0	0
June 23	0.3	24.4	22.7	3.4
June 27	1.5	48.8	12.4	10.4

Table 2. Evolution of THMFP Along the Treatment Line

Sample THM Date (ppb)	Raw Water	Filtered Water	Ozonated Water	GAC-Filtered Water
April 10	44.4	31.9	25.6	3.4
May 19	145.9	69.8	114.2	14.5
May 26	121.1	66.2	80.9	6.1
June 23	207	90	89	19
June 27	132.8	58.4	91.3	16.8

ozonation does not remove THMs. GAC filtration is efficient during the first 3 months, after which breakthrough occurs. During the duration of the study, THMFP of the raw water was between 44 and 207 μg/L (see Table 2). A 30 to 60% THMFP removal is obtained by prechlorination and clarification. Under these conditions, ozonation does not affect the THMFP. THMFP levels decrease dramatically after ozone-GAC treatment.

CLSA-GC/MS

Table 3 presents the compounds identified by GC/MS and their relative concentrations. In the clarified water, 42 compounds are present. Among them, several could be responsible for tastes and odors, such as terpenes, terpenol, sesquiterpenes, and brominated and iodinated THMs formed during chlorination of water containing iodide and bromide. The range of micropollutants concentration is between 1 and 1350 ng/L. Ozonation and GAC filtration remove many of these micropollutants, particularly the taste- and odor-causing compounds. Only nine compounds are identified after GAC filtration. One example of the CLSA chromatogram of raw water is presented in Figure 6.

Flavor Profile Analysis (FPA)

Several taste and odor analyses were performed. The average results of FPA are listed on Table 4. The results show that taste and odor removal by ozonation

Table 3. Evolution of the Nature and Concentration of Compounds Identified on Water Along the Treatment Line

Name of Organic	Quantity (ng/L)		
	Sand-Filtered Water	Ozonated Water	GAC-Filtered Water
Alkane C7	240	400	—
Dichlorobromomethane	150	150	—
Hexanal	—	15	—
Toluene	1350	620	150
Dibromochloromethane	160	200	—
Dimethylhexane	65	55	—
Ethylbenzene	50	40	—
Xylene	75	130	—
Dichloroiodomethane	40	—	—
Alkylbenzene	240	80	—
Styrene	45	25	—
Terpene	40	20	—
Alkane C9	+	100	—
Bromoform	100	100	—
Haptanal	+	150	—
Alkane C10	20	50	—
Bromochloroiodomethane	+	—	—
Alkylbenzene	25	—	—
Methylstyrene	30	—	—
Octane	—	40	—
Alcohol C8	70	360	150
Dimethylstyrene	25	—	—
Nonane	—	75	—
Terpinol	+	—	—
Decane	20	50	—
Alkane C14	25	—	—
Alkane C15	25	—	—
2,6-Di-*tert*-butylquinone	430	150	200
Sesquiterpene	20	—	—
Alcohol C13	25	15	—
1,5 Di-*tert*-butyl-3,3-dimethylbicyclo-(3,1,0)-hexanone	175	45	50
Trimethylpentane-1,3-diisobutyrate	125	65	250
Dibutylthiophene	+	—	+
Alcohol C14	10	—	—
Alkane C18	30	—	—
4-Ethyl-2,6-di-*tert*-butylphenol	510	65	150
Nonylphenol	+	—	—
Alcohol C16	10	—	—
4-Isocyano-2,6-di-*tert*-butylphenol	+	—	—
Alkane C19	15	—	—
Alkane C20	15	—	—
Dibutylphthalate	25	20	30
Alkane C21	+	—	—
Alkane C22	+	—	—
Dioctylphthalate	45	25	140

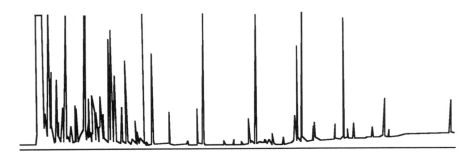

FIGURE 6. Example of CLSA-chromatogram obtained on Macau raw water.

Table 4. Evolution of the Nature and Intensity of Tastes and Odors Along the Treatment Line

Samples	Taste and Odor	Intensity
Raw water	Musty, earthy	10
Sand-filtered water	Musty	8
Ozonated water	Musty	2
	Astringent	2
GAC-filtered water	Astringent	2
Sand-filtered water after ozonation	Musty	2
	Astringent	2

and GAC filtration is very efficient. In the clarified water, the earthy and musty tastes are very strong, the intensity is equal to 8. After ozonation, the musty taste level decreases significantly, with intensity less than 2. After GAC filtration, the musty taste is removed, leaving a trace of astringent taste.

CONCLUSION

This study evaluated the efficiency of the combination of ozonation with GAC filtration in order to increase the overall quality of Macau drinking water. Results have shown that the organic content of the clarified water is quite low (TOC = 1.5 mg/L). Numerous compounds (42) were identified at a concentration range of 1 to 1350 ng/L and many could be responsible for tastes and odors (terpenic and iodinated compounds). Musty and earthy tastes and odors are most predominant.

Macau water can be effectively treated by ozonation and GAC filtration to reduce THMs, micropollutants, and tastes and odors. In addition, this combination is effective against detected or undetected accidental pollution which could occur in this new industrial area. Pilot-plant studies will continue to be used to evaluate ozonation of the settled water, followed by a first-stage GAC filtration.

At the same time, kinetic and stoichiometric data are produced at different periods of the year in order to optimize design of a new ozonation plant.

ACKNOWLEDGMENTS

The authors acknowledge the assistance of other members of Laboratory of Macau Water Supply Co. Ltd. — U. Sam Chan, Chan Kin Shun, and Zhang Guon Rong — and Laboratoire Central de la Lyonnaise des Eaux/Dumez — Joël Mallevialle and Jean Pierre Duguet.

REFERENCES

1. Wable, O., Duguet, J. P., Brodard, E., Mallevialle, J., and Roustan, M. "Computer Aided Decision in Ozone Contactor Design and Operation: An Expert System," paper presented to the 9th I.O.A. Ozone World Congress, New York, NY, June 3 to 9, 1989.
2. Maier, D., and Kurzmann, G. E. "The Determination of Higher Ozone Concentrations," *Wasser Luft Betr.* 21:125-128 (1977).
3. Maier, D. "Methods for the Determination of Ozone in Concentrated Gas Mixtures: in *Ozonation Manual for Water and Wastewater Treatment,* W. J. Masschelein, Ed. (New York: John Wiley & Sons, 1982), pp. 151-153.
4. Cohen, H. "Comparison of Methods for the Determination of Ozone in Aqueous Solution", MS Thesis, Miami University, Oxford, OH (1981).
5. Bader, H., and Hoigne, J. "Determination of Ozone in Water by the Indigo Method: A Submitted Standard Method," *Ozone Sci. Eng.* 4:169-176 (1982).
6. Bader, H., and Hoigne, J. "Analysis of Ozone in Water and Wastewater by an Indigo Method," paper presented at 4th Ozone World Congress. I.O.A., Houston, TX, November 1979.
7. Duguet, J. P., Brodard, E., Roustan, M., and Mallevialle, J. "The Development of an Automated Procedure and the Applicability of this Procedure for Monitoring the Effectiveness of Ozone," *Ozone Sci. Eng.* 8:321-338 (1986).
8. "Strategies for the Control of Trihalomethanes," Paper #20174, AWWA Seminar Proceedings, AWWA Conference, June 5, 1983.
9. "Identification and Treatment of Tastes and Odors in Drinking Water," Cooperative Research Report, J. Mallevialle and I. H. Suffet, Eds., AWWARF (1987).
10. "Standard Methods for the Examination of Water and Wastewater," 15th ed., APHA, AWWA, WPCF (1981).
11. Chrostowski, P., Dietrich, A. M., and Suffet, I. H. "Laboratory Testing of Ozonation Systems Prior to Pilot Plant Operations," *J. Am. Water Works Assoc.* 74:38 (1982).

CHAPTER 16

A Pilot-Plant Study on Advanced Treatment of Potable Water

W. Dazhl and L. Bingjie

INTRODUCTION

The Song-Hua River is the water supply source in Harbin. The 7th Water Works of the Harbin Water Supply Company was designed by the northeast branch of China Municipal Engineering Design Institute in 1975, with a capacity of 160,000 m^3/day, of which 67,000 m^3/day is for potable use. The total capital investment was more than 30 million Yuan (RMB). Previously, potable water was not supplied from this river for as long as 14 years because the Song-Hua River is heavily polluted by industrial and domestic wastewater. In dry seasons, the average chemical oxygen demand (COD) of the river was as high as 48 mg/L with total organic carbon (TOC) of 11.7 mg/L. Carcinogens such as benzpyrene(a), chloroform, carbon tetrachloride, chlorobenzene, BHC, DDT, nitrobenzene, arsenic, cadmium, and chromium were found. Pollution by phenols was the heaviest, with the values in dry seasons as high as 0.032 mg/L, which is 16 times higher than the allowable amount fixed by the National Water Quality Standard of the Peoples Republic of China. The total number of bacteria and coliform also exceeded the National Standard for Source Water. The concentrations of ammonia, nitrites, and nitrates were also high. The results showed that this source was one of the most polluted water sources in the world. The Petro-Chemical Research Institute and Analytical Center of Heilong jiang province have found 10 carcinogenic substances among 38 organic chemicals that were found in the water during the period of 1980 to 1982. Of these organic compounds, 17 are priority pollutants listed by the U.S. EPA. Harbin University of Medicine reported that the water in the Song-Hua River showed positive Ames test response from concentrates of 3 L of raw water.

From 1985 to 1987, a 1000 m^3/day pilot plant was developed. Three alternative treatment methods were chosen for research (1) moving prechlorination out of the conventional purification process, moderately increasing the dosage of coagulant and controlling parameters in sedimentation and filtration, making the filter effluent lower than 2° in turbidity, i.e., improving the conventional process; (2) after these processes were used, ozonation was added; and (3) after improving the total process, ozonation-activated carbon adsorption was added as final steps before final chlorination.

The 2-year study showed that the polluted raw water, after prechlorination, has a larger trihalomethane (THM) value and an increase in the conductivity of drinking water. Prechlorination should be eliminated to reduce formation of halo-organics. Polluted raw water after conventional treatment and followed by ozonation removed some organic pollutants, but toxic by-products may be formed in the disinfection process. The same raw water, after conventional treatment and ozonation-activated carbon adsorption, may also have some toxic products formed after final chlorination. When the water is treated by subsequent activated carbon, the conductivity of the water is reduced. Furthermore, activated carbon is effective for removal of many pollutants such as COD, color, taste, odor, phenol, nitrobenzene, lignin, oil, chloroform, ammonia, nitrite, nitrate, BHC, DDT, organic chlorine, cyanide, manganese, iron, arsenic, cadmium, chromium, selenium, mercury, etc.

This pilot-plant study included engineering changes planned for the more economical use of the 7th Water Works in Harbin. Treatment plants with similar water sources may find this study of alternative treatment methods useful to understand.

EXPERIMENTAL PROCESS AND PARAMETERS STUDIED

Three schemes were chosen for experimental research:

1. The conventional purification process without prechlorination. A moderate increase dosage of coagulant was added and control parameters for efficient coagulation-sedimentation-filtration were properly chosen to keep the turbidity of filter effluent lower than 2°. This is called the intensified and improved conventional purification process.
2. Intensified and improved process followed by ozonation.
3. Intensified and improved process followed by ozonation-activated carbon adsorption. This is shown in Figure 1.

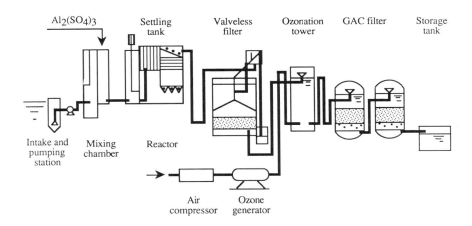

FIGURE 1.

In the first scheme, the existent coagulation-sedimentation-filtration system which has a capacity of 6000 m³/day and an additional installed ozonation-activated carbon system with the capacity of 750 to 1000 m³/day were used for the experiment.

During coagulation-sedimentation, aluminum sulfate dosages were increased from 40 to 80 mg/L. In order to increase the effectiveness of flocculation, four stirrers were used in rotary baffled flocculation tanks. Reaction took place under varied velocities in the range of 0.5 to 0.2 m/sec and reaction times of 20 to 23 min. In the upward flow tube settling tank, the planar upward velocity was 1 to 2 mm/sec. In the gravitational valveless filter, double layers of media were used; gravel diameters of quartz sand and anthracite were 0.5 to 1.0 mm and 1.0 to 1.8 mm, respectively; the thickness of both layers was 0.4 m and the filtration rate was 8 to 10 m/hr.

Two sets of XY-55 ozone generators were used in the ozonation system with a working pressure of 0.4 to 0.8 kg/cm², adjustable voltage 8 to 15 kv, and output flow of 250 to 280 g/hr. Filter effluent entered the ozonation contact tower from the top and the ozone concentration was about 2% by weight. Ozonated air entered the tower from the bottom and passed through microdiffusers and came into contact with the water by countercurrent flow to make mixing and oxidation more efficient. Operation parameters were: tower height, 5.5 m; diameter, 600 mm; flow velocity of water in the tower, 25 m/hr; contact time, about 12 min; and ozone dose, 2 to 4 g/m³.

The ozonation tower effluent was sent to two granular activated carbon (GAC) filters in series. Each filter was 2 m in diameter and 5 m in height. The thickness of the filter bed of the first filter was 1200 mm and the second one was 1800 mm. Both filters were filled with ZJ-15 GAC. Filtration rates were controlled in the range of 7 to 10 m/hr. The empty bed contact time was 18 to 26 min.

Table 1. Pollutants Removal

Parameter	Raw Water	Filter Effluent		Effluent of Ozonation		Effluent of GAC	
		Average	Removal %	Average	Removal %	Average	Removal %
COD (mg/L)	39.84	20.15	50	11.952	70	5.976	85
UV	0.289	0.145	50	0.072	75	0.035	88
Color (°)	64	27	58	13	80	5	92
Phenol (mg/L)	0.010	0.008	20	0.003	70	0.001	90
Nitrobenzol (μg/L)	0.151	0.130	14	0.082	45	0.060	60
Lignin (mg/L)	1.350	1.120	17	0.675	50	0.270	80
Oil (mg/L)	0.750	0.660	12	0.150	80	0.060	92
Chloroform (μg/L)	19	62				5	74
NH_3-N (mg/L)	1.053	0.995	6	0.864	18	0.600	43
NO_2-N (mg/L)	0.073	0.059	19	0.024	67	0.007	90
NO_3-N (mg/L)	0.231	0.214	7	0.184	20	0.023	90
BHC (μg/L)	0.390	0.287	31	0.230	41	0.167	57
Organic chloride (μg/L)	0.251	0.226	10	0.168	33	0.114	55
Cyanide (mg/L)	0.028	0.019	32	0.002	93	(—)	100
Manganese (mg/L)	0.154	0.098	36	0.081	47	0.046	70

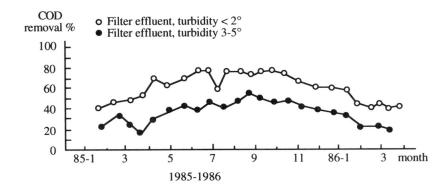

FIGURE 2. COD removal by filtration at two different influent turbidity levels.

The backwash rate was 25 m/hr, duration about 15 min, and backwash was completed every 5 to 8 weeks.

RESULTS

The water quality and flow of the Song-Hua River fluctuated considerably because of seasonal variation. In flood seasons, turbidity was very high, but the concentrations of pollutants were lower than in the dry seasons. In dry seasons, the turbidity was low, but the pollution concentration increased 10-fold as compared to the flood seasons. During the study period, the effluents of filter, ozonation tower, and GAC filter were sampled. The results are summarized in the following sections.

Removal of Pollutants

Table 1 shows the pollutants removed by filtration, ozonation, and GAC.

COD Removal

Figure 2 shows that better control of the turbidity of the purified water is important for efficient COD removal. Figure 3 shows that COD removal by the ozonation process is about 70%; sometimes COD values of the process effluent are higher than 10 mg/L. COD removal by the GAC filter is about 85%, with the effluent values all below 10 mg/L.

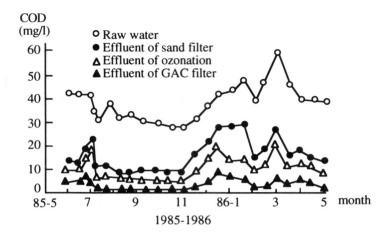

FIGURE 3. COD removal by unit operations of filtration, ozonation, and GAC.

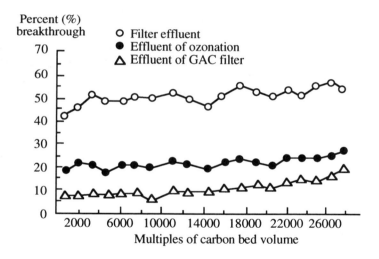

FIGURE 4. Percent (%) breakthrough from a GAC bed vs bed volumes of water.

Removal of Ultraviolet Absorbing Compounds

Ultraviolet absorbance reflects organic pollution levels in the water and the efficiency of pollutant removal. Figure 4 shows UV breakthrough curves. If UV breakthrough is 12%, then COD removal will be 88%. A consistent pattern exists with COD removal.

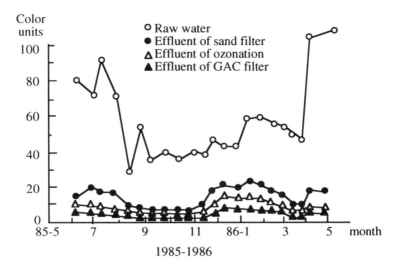

FIGURE 5. Color unit removal by unit operations of filtration, ozonation, and GAC.

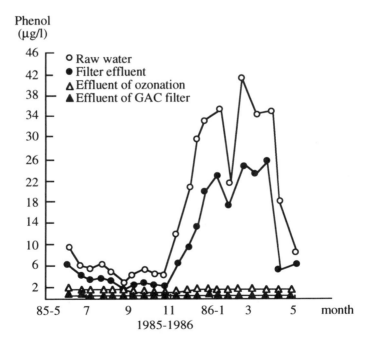

FIGURE 6. Phenol removal by unit operations of filtration, ozonation, and GAC.

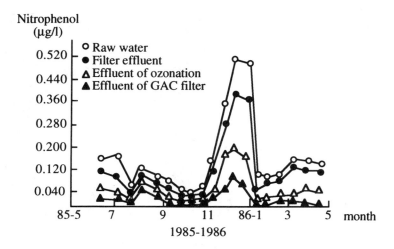

FIGURE 7. Phenol removal by unit operations of filtration, ozonation, and GAC.

Removal of Color, Phenol, Nitrophenol, Lignin, and Oil

These pollutants are not significantly affected by conventional purification processes. The National Standard for Potable Water can only be achieved by using ozonation-GAC treatment (see Figures 5 to 9).

Chloroform Removal

Chloroform is a well-recognized carcinogen. In other countries, THMs are usually used as a controlling standard for potable water. Our new Peoples Republic of China National Standard stipulates that the chloroform content should be lower than 60 μg/L. Because this river is heavily polluted, using prechlorination before the THM formation potential (THMFP) is removed causes the THM value in the filter effluent to increase. During our experiments the dosages of chlorine were 4 to 10 mg/L; residual chloride in filter effluent was 1.0 to 2.0 mg/L. Chloroform concentration increased to more than 100 μg/L. (see Figure 10). Conventional prechlorination processes followed by ozonation-GAC can remove THM, but the service life of GAC is very short. Elimination of prechlorination and using conventional purification followed by advanced treatment can remove THMFP efficiently. A final treatment by moderate chlorination of the GAC filter's effluent was used. From Figure 10, it is observed that the chloroform content is very low and is lower than the standard. Generally, the chlorine dosage is controlled in the range of 0.5 to 0.7 mg/L.

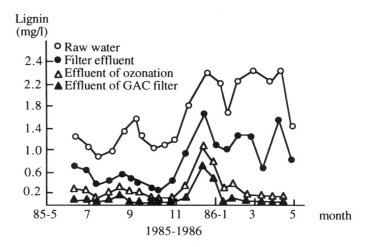

FIGURE 8. Lignin removal by unit operations of filtration, ozonation, and GAC.

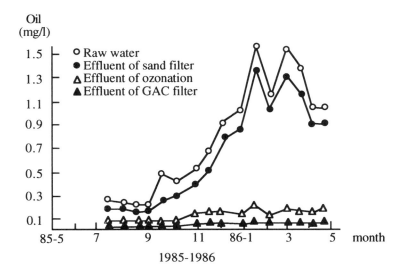

FIGURE 9. Oil removal by unit operations of filtration, ozonation, and GAC.

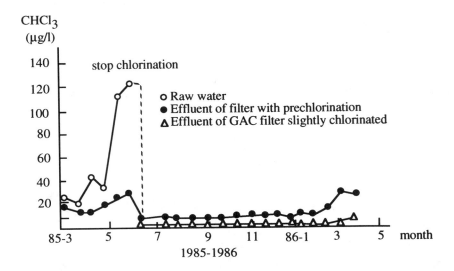

FIGURE 10. Chloroform removal with different unit operations.

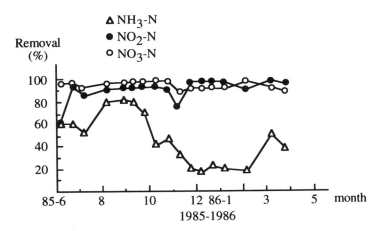

FIGURE 11. The curve of nitrogen removal by ozonation-GAC filter.

Removal of Ammonia and Nitrate

Figure 11 shows the ammonia, nitrite, and nitrate concentrations in the effluents after treatment by ozonation-GAC. As shown, from June to October, removal of ammonia is higher than 60%, but the removal is only about 25% for the remainder of the year. This indicates biological activity occurring on the

Table 2. Total Bacteria Number and Coliform Colony Test Result

Parameter	Raw Water	Effluent of Ozonation	Effluent of GAC Filter
Total bacteria (pec/mL)	629,500	1	20
Coliform colony (pec/L)	2,858,000	<3	<3

carbon column is related to water temperature. At temperatures higher than 10°C, ammonia removal is good. Removal is poor at temperatures lower than 5°C. Figure 11 also shows that removal of nitrite is about 90% and the removal of nitrate exceeds 90%.

Removal of Other Pollutants

Each of the three treatment schemes removes some BHC, organic chlorine, cyanide, and manganese. Ozonation-GAC significantly removes iron, copper, aluminum, zinc, chromium, cadmium, silicon, and mercury.

Disinfection by Ozone

To kill pathogens and viruses, ozone is more efficient than chlorine. Experimental results show that after ozone treatment in heavily polluted raw water, the total number of bacteria and coliform are below the limit set by the National Standard for Potable Water. After GAC filtration, bacterial numbers increased considerably, but coliform remains stable (see Table 2).

Ames Test

During the dry season in 1985, Ames tests of raw water and effluent treated by prechlorination and the conventional purification processes of the 7th Water Works were completed. Table 3 shows the Ames test results of treatment test with TA_{98} without adding the S-9 activator for 2-L raw water samples. The number of colonies in each test did not surpass two times that of the natural bacterial colony. Ames testing of 0.32 L effluent, treated by prechlorination and conventional purification processes indicated positive results. When 0.064 L were used, chromosome exchange also indicated a positive. This shows that mutability distinctly increases.

During the dry season in 1986, Ames tests for raw water and effluents of each unit of advanced treatment were completed (see Table 4). The results of the TA_{98} test without adding the S-9 activator showed that for 3 L of raw water,

Table 3. Ames Test Results (Colonies per Plate)

Sample	Concentration (L/plate)	−S-9		+S-9	
		TA_{98}	TA_{100}	TA_{98}	TA_{100}
Raw water	0.13	15	109	25	85
	0.32	22	106	25	106
	0.8	19	112	30	104
	2.0	30	145	28	105
Purified water	0.13	21	127	21	98
	0.32	54	150	25	88
	0.8	62	234	27	119
	2.0	98	316	56	184
Natural rechange		25	124	18	104
Dioxan	62.5 (μg/plate)	1145	1324		
2-Amino fluorene	10 (μg/plate)			1100	1343

Table 4. Ames Test Results (Colonies per Plate)

Sample	Concentration (L/plate)	−S-9		+S-9	
		TA_{98}	TA_{100}	TA_{98}	TA_{100}
Raw water	1.0	26	171	37	122
	2.0	24	216	39	96
	3.0	33	178		
Effluent of sand filter	0.1	27	112	33	80
	0.3	22	116	33	105
	1.0	26	72	53	49
	3.0	29	Little colonies	28	102
Effluent of ozonation	0.1	24	100	30	99
	0.3	29	113	37	111
	1.0	38	79	47	78
	3.0	65	66	91	109
Effluent of GAC filter	0.1	26	87	29	96
	0.3	28	89	32	96
	1.0	27	86	29	84
	3.0	33	76	38	95
After disinfection by chlorination	0.1	24	206	28	159
	0.3	22	191	26	124
	1.0	23	187	33	120
	3.0	23	156	31	115
Natural rechange		24	136	28	144
Dioxan	62.5 (μg/L)	1477	1011		
2-Amino Fluorene	10 (μg/L)			944	952

the Ames test results were negative, 0.09 L of raw water used for chromosome exchange also indicated a negative, and a 3 L effluent of a sand filter also was negative. A 3-L effluent of ozonation surpassed two times the number of natural bacterial colony, indicating a positive Ames test, whereas GAC filter effluent was negative. Adding chlorine after GAC filtration also showed a negative Ames

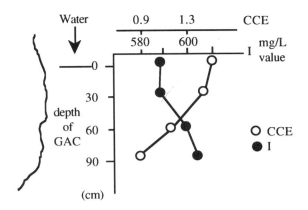

FIGURE 12. I and CCE values at different depths of GAC bed.

test result, which was confirmed by a 0.4-L chromosome exchange test. Thus, after ozonation there are some mutagens harmful to human health in the product water, but after GAC adsorption, mutability is considerably reduced.

The GAC filter was operated for over 1 year. During this period, the total amount of filter effluent corresponded to 28,000 bed volumes. Samples of carbon were taken at different depths of the GAC bed at the end of the run. Iodine value and chloroform extract of the column were completed. (see Figure 12). It is calculated that the GAC filter can be put into continuous operation for as long as 3 years and the total amount of water to be treated can reach 790,000 m^3. Thereby, it is estimated that the cost of water treated by ozonation-GAC increases $0.03 for each cubic meter.

CONCLUSIONS

1. Moderately increasing coagulant doses, properly controlling parameters in conventional purification processes of flocculation-sedimentation-filtration, and keeping turbidity of filter effluent lower than 2° are effective for pollutant removal. Treating polluted raw water by prechlorination produces Ames positive results in potable water. Prechlorination should be eliminated to reduce THM formation. This optimal conventional purification scheme can economically treat severely polluted river water.
2. Conventional treatment followed by ozonation can remove pollutants, but reaction products are formed, some of which are mutagens. Subsequent GAC filtration reduces the mutagens.
3. The polluted raw water of the 7th Water Works, after treatment by an optimized conventional purification process and ozonation, still

could not meet all requirements of the new Peoples Republic of China's National Standard for Potable Water. Treatment by ozonation-GAC filtration significantly removes various pollutants from water, such as COD, color, taste, odor, phenol, nitrobenzene, lignin, oil, chloroform, ammonia, nitrite, nitrate, BHC, DDT, organic chlorine, cyanide, manganese, iron, arsenic, cadmium, chromium, selenium, mercury, etc.
4. This pilot plant has effectively provided both technical and economical parameters for the optimization of the operation of the 7th Water Works. The information is valuable for advanced treatment of similar water sources in the Harbin water supply.

CHAPTER 17

Ozonation of Organic Compounds Causing Taste and Odor Problems

C. Anselme, J. P. Duguet, J. Mallevialle, and I. H. Suffet

ABSTRACT

This paper will present a short literature review on the efficiency of ozonation on taste and odor removal. A case history of the use of the flavor profile analysis (FPA) method is presented to help evaluate the use of the oxidant ozone at a water treatment plant. A correlation between the flavor profile analysis method and chemical analysis is done to try to understand how the oxidant is removing taste and odor compounds and thereby enabling control over organoleptic problems.

INTRODUCTION

The periodic occurrence of objectionable tastes and odors has plagued the water supply industry for many years. Taste and odor (organoleptic) problems are associated with unsafe water by the public regardless of the actual quality of the water. This leads to customer dissatisfaction and bad public relations. In addition, customers may choose alternate untreated water sources which may be hazardous to their health.[1]

The importance of the organoleptic properties of water — its appearance, taste, and odor — is recognized worldwide, and many countries have some type of sensory water quality standard. The World Health Organization,[2] for example, stipulates that neither the taste nor odor of water shall be objectionable to 90% of the populace. The U.S. Public Health Service's (PHS) drinking water standards stipulated in 1925 that the taste of drinking water be generally acceptable; in 1946, it stipulated that drinking water be "not objectionable;" and in 1962, "not offensive."[3,4]

The U.S. PHS[4] regulations and, more recently, the National Secondary Drinking Water Regulations[30] of 1979, promulgated by the U.S. Environmental Protection Agency (1979), quantified the odor standard by stipulating that the threshold odor number (TON-the number of times a water must be diluted with odor-free water before the odor is just barely perceptible) should be 3 or less to ensure public acceptance. The Council of European Communities[5] recommends "dilution numbers" (instead of TONs) of 0 for both taste and odor, but specifies maximum admissible dilutions of 2 at 12°C and 3 at 25°C for both taste and odor.

Although Standard Methods[6] recommends panels of not less than five members at many water utilities, one plant control person usually completes the test alone. Furthermore, the results do not include a description of the quality of the perceived taste and odor.

Description of an objectionable taste or odor is subjective and depends on the individual's background as well as training. Taste has four basic qualities: sweet, sour, salt, and bitter. There are 30 to 40 primary odors.[7] The nose and palate are physically connected; therefore, the simultaneous detection of any combination of the four taste qualities and any of the numerous odor qualities produces the flavor associated with a food or beverage.

In the work presented in this paper, the FPA method has been used.[8] The FPA technique was developed for the food industry[9] and has been recently extended to the analysis of tastes and odors in water samples.[10] Erlenmeyer flasks and a temperature of 45°C were adopted for odor evaluations.[11,12] Batch samples of water were collected, headspace free, and kept refrigerated until analysis. A trained panel of four or more persons evaluated the organoleptic qualities (both flavor and aroma), giving a description (e.g., musty, fruity, astringent) and an intensity value for each taste and odor sensation observed. The intensity value is recorded as the average of the intensities as perceived by at least 50% of the panelists for a particular descriptor. The seven-point scale for intensity is 1 (threshold), 2, 4, 6, 8, 10, 12. Common odors observed in raw drinking water are shown in Table 1. Common tastes observed in treated drinking water are shown in Table 2. Raw drinking water is never tasted because of potential microbial contamination.

Table 1. Common Odors in the Influent of Three Water Utilities

Sample Location	Major Response	Frequency of Occurrence	Intensity
Philadelphia	Sewage	83%	4.3
Suburban	Creeky	20%	3.5
Water Co.	Musty	10%	3.7
Philadelphia	Decaying Vegetables	62%	5.0
Water	Septic	52%	4.0
Department	Vegetation	22%	4.3
	Earthy	43%	4.4
	Musty	17%	3.6
	Fishy	13%	5.0
Lyonnaise des	Muddy	71%	6.2
Eaux	Fishy	71%	6.5
	Musty	38%	3.5
	Septic	29%	9.0

Note: Flavor profile intensity scale = 1 (Threshold), 2, 4, 6, 8, 10, 12; seven-point scale.

Reprinted from J. AWWA, Vol. 88, No. 10 (October 1988). By permission of American Water Works Association.

Table 2. Common Tastes in the Effluent of Three Water Utilities

Sample Location	Major Response	Frequency of Occurrence	Intensity
Philadelphia	Chlorinous	72%	2.2
Suburban	Metallic	38%	2.1
Water Co.	Astringent	28%	2.0
	Bitter	22%	2.7
Philadelphia	Chlorinous	54%	2.7
Water	Musty	37%	2.2
Department	Earthy	19%	2.3
Lyonnaise des	Chlorinous	94%	4.6
Eaux	Musty	18%	2.2
	Astringent	18%	2.2

Note: Flavor profile intensity scale = 1 (Threshold), 2, 4, 6, 8, 10, 12; seven-point scale.

Reprinted from J. AWWA, Vol. 88, No. 10 (October 1988). By permission of American Water Works Association.

LITERATURE REVIEW — WATER TREATMENT FOR REMOVAL OF TASTE AND ODOR COMPOUNDS

The effectiveness of conventional water treatment processes without the use of activated carbon to remove organoleptic compounds was reviewed to determine the most effective procedures to solve taste and odor problems. The processes evaluated were coagulation, filtration, and oxidation (chlorine, chlorine dioxide, chloramines, potassium permanganate, and ozone).[13,33] Laboratory experiments were also performed on a group of over 15 compounds including known organoleptic compounds such as geosmin and methyl isoborneol.[13] Generally, the organoleptic compounds tested were resistant to coagulation, filtration, and, in most cases, oxidant treatment by chlorine dioxide, chloramines, and potassium permanganate.

In general, the degree of degradation of an organic compound by an oxidant depends upon many factors including (1) the strength of the oxidant, (2) the way the oxidant attacks the bonds of the compound, (3) the structure of the compounds, and (4) environmental factors such as pH, temperature, concentration of chemicals, and presence of interfering substances. The literature and laboratory experiments suggest that the chlorine-type oxidants and potassium permanganate are not an effective way to remove most low molecular weight organic compounds. Nevertheless, some water treatment plants have had good results in solving specific taste and odor problems using oxidants. Oxidants, therefore, may be good for specific events.

At present, when an odor event occurs, it is difficult to identify the responsible agents and thus determine if an oxidation process would be effective. Even if the agent is known, it is difficult to predict its response to the oxidant because of the many factors involved in the reaction. Therefore, water treatment plants try to solve taste and odor problems with oxidants by trial and error with immediate evaluation of the results. The oxidant ozone was not evaluated in this laboratory study: however, it was tested in a field study[29] at the Morsang Water Treatment Plant in France and is described here.

A literature review of the different mechanisms of the action of ozone on aqueous organic pollutants by Dore,[14] based upon the work of Hoigne and coworkers,[15-21] indicates that ozone can react directly to form carbonyls by reacting as either:

1. A dipole on C=C double bonds
2. An electrophilic agent on aromatics by ring hydroxylation
3. A nucleophilic agent on C=N double bonds

Ozone can also react indirectly as a free radical to form carbonyls.

Scavengers such as carbonate can stop the reaction. During water treatment, intermediate reaction products can be formed. The type and amount of products are a function of ozone dose, reaction time, scavengers, and pH.

Direct ozone reaction rate constants have been determined for a wide range of organics at pH 2 to 3.[22] The rate laws are assumed to be first order with respect to ozone (O_3) and solute concentration (M). Therefore, the action can be expressed as a second-order reaction.

$$M + nO_3 \xrightarrow{k_{O_3}} \text{Products} \qquad (1)$$

$$\frac{-d(M)}{dt} = \frac{k_{O_3}}{n} (O_3)^{1.0} (M)^{1.0} \qquad (2)$$

Hoigne and Bader[22] proposed the following relationship for compounds treated with ozone under water treatment plant conditions:

$$t_{1/2} = \frac{0.69}{(O_3)k_{O_3}/n} = \frac{0.69}{10^{-5}k_{O_3}/n} \tag{3}$$

This follows mathematically from the second-order rate equation. The right-hand approximation can be made assuming 10^{-5} M ozone (0.5 mg/L, typical in treatment) and that n = 1.

Using this approximation, compounds having rate constants greater than 100 M^{-1} sec^{-1} are significantly oxidized in 10 min by the direct attack mechanism. Hoigne and Bader suggest that those compounds would be oxidized by ozone under water treatment plant conditions at low pH. At higher pHs, where the hydroxyl free radical predominates Equations 1 and 2 hold with O* (representing any oxygen radical like the hydroxy radical) replacing O_3:

$$M + O^{\cdot} \xrightarrow{k'} \text{Products} \tag{4}$$

$$\frac{-d(M)}{dt} = k'(O^{\cdot})^{1.0} (M)^{1.0} \tag{5}$$

Staechelin and Hoigne[23] conclude that in the case of impure water (e.g., water treatment), the overall kinetics due to radical-type chain reactions may still be first order in ozone concentration if the concentration of solutes stays constant. However, as a result of a sum of chain reactions of each solute with the free radicals, each solution concentration is affected differently. For these reasons, prediction of products at present has not been accomplished.

CASE STUDY — LYONNAISE DES EAUX, MORSANG WATER TREATMENT PLANT

Conventional Treatment of Taste and Odors

Figure 1 shows the Morsang Water Treatment Plant near Paris, France. River Seine water enters the plant and is split to Line 1 and Line 2. Breakpoint chlorination with 1.5 to 3.0 mg/L chlorine is used before the water flow is split. The ammonia level in the river is 0.1 to 0.2 mg/L, thus free residual chlorine is present in the process. On Line 1, normally 5 to 15 g of powdered activated carbon is injected per cubic meter of water for taste and odor control. Between May and July, up to 25 g of powdered activated carbon can be applied. The water in Line 1 then enters a coagulation process (Pulsator-Degremont, France).

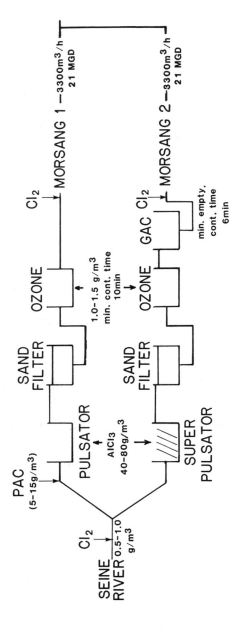

FIGURE 1. Morsang water treatment plant. (Reprinted from *J. AWWA*, Vol. 88, No. 10 [October 1988]. By permission of American Water Works Association.)

The prechlorinated water in Line 2 is passed directly to a Super-Pulsator, where alum (40 to 80 g of alum per cubic meter water) and activated silica are utilized for coagulation. Each line then follows with rapid sand filtration. The chlorine residual after sand filtration is <0.1 mg/L. An ozone dosage of 2.5 to 2.7 mg/L is then applied. After a 10-min contact time, an ozone residual of 0.2 leaves the ozonation process. All the odors found in raw water influents decrease in percent of frequency of occurrence and intensity except chlorine and musty-earthy odors during the treatment process before ozonation. The increase in the percent frequency of occurrence of musty and earthy odors from raw water to prechlorinated water indicates that the prechlorination process of conventional treatment is responsible for the increase. An increase of contact time during the settling and sand filtration process usually increased the frequency of occurrence. After sand filtration, the residual chlorine is 0.1 to 0.2 mg/L as Cl.

Ozonation Treatment

The chlorine residual after sand filtration is <0.1 mg/L. An ozone dosage of 2.5 to 2.7 mg/L is then applied and after a contact time of 10 min, an ozone residual of <0.2 mg/L leaves the ozonation process. Thus, during the 10-min process of ozonation on Morsang Lines 1 and 2, residual chlorine will react with the ozone and produce chloride (77%) and chlorite (23%) according to Haag and Hoigne.[22] This reaction is pH-dependent. Haag and Hoigne[22] concluded that under water treatment practice, e.g., pH = 8, the rate of reaction is comparable to the rate of ozone consumption and faster than chlorine consumption by other solutes present in the water. Thus, during ozonation, chlorite and ozone oxidation of organic solutes are possible. The ozone residual of 0.1 to 0.2 mg/L leaving this process will be reduced to oxygen by passage through the subsequent granular activated carbon (GAC) process on Morsang Line 2.

Ozonation is a very effective unit process for taste and odor removal in this treatment train. All the odors present in the raw water (see Table 1) are reduced to less than 50% frequency of occurrence by ozonation. Fishy tastes and odors are completely eliminated by ozone treatment. Fruity odors, which were produced during ozonation, increased to 40%. Figures 2 through 6 present intensity values from the flavor profile analysis over the time period of the ozonation process on Morsang Line 2. The intensities of all descriptors that are listed in Table 1 for Lyonnaise des Eaux (except for an astringent taste sensation) are reduced to below the slight level on the flavor profile analysis scale (intensity 4.5) and, as noted earlier, only the fruity odor increases. The occurrence of earthy, musty, and muddy tastes and odors decreases from 30 to 50% after ozonation. In both Morsang Lines 1 and 2, ozonation decreases muddy, musty, and earthy odors to less than 40% occurrence (muddy taste) and less than 25% occurrence (earthy, musty, and chlorine taste and odor). The musty odor in Morsang Line 2 was

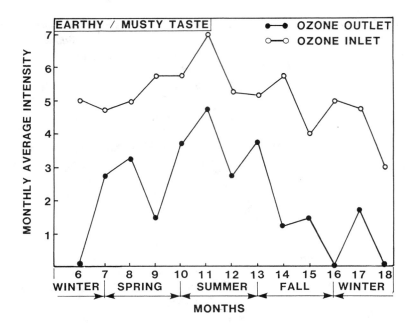

FIGURE 2. (Reprinted from *J. AWWA*, Vol. 88, No. 10 [October 1988]. By permission of American Water Works Association.)

different from that in Line 1. Also, the earthy taste has a 50% decrease in occurrence.

The development of fruity, fragrant, and orange-like odors with the ozonation process has been noted (see Figure 6). Correlations between oxidation treatment and these descriptors is significant for both lines of treatment at the Morsang plant. The fruity odor occurrence increased from 5 to 35% after ozonation and decreased to 22% after GAC filtration.

Figure 6 shows that the fruity odor increased from a negligible intensity to just perceptible in the winter; increased in spring, summer, and into the fall to about a monthly average of 2 and then dramatically increased to a high of 4 to 7 in subsequent months. The increase in fruity odors is suspected to be due to the formation of aldehydes during ozonation. This is a hypothesis that was tested using the statistical correlation work described later in this paper.[24]

Astringent tastes (see Figure 4) were not consistently removed by ozone through the 18 months of the study. At some times, ozone did decrease astringency but at other times the opposite occurred. In any case, slight odors of 1 to 2 FPA intensity change were usually involved. Plastic and pharmaceutical tastes (see Figure 5) were mostly at 1 to 2 FPA intensity entering the ozonation process.

In summary, the data presented from Morsang shows how the background influent odors change during treatment. Table 3 summarizes the taste and odor

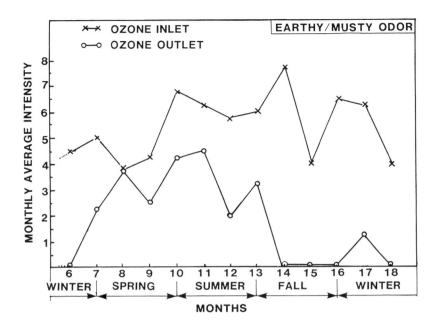

FIGURE 3. (Reprinted from *J. AWWA*, Vol. 88, No. 10 [October 1988]. By permission of American Water Works Association.)

changes caused by ozonation on Morsang Lines 1 and 2. Two interesting points were observed. First, earthy and musty tastes and odors increase in frequency of detection as the water progresses through prechlorination and settling. Most other qualities decrease in frequency of detection. The intensities of all qualities show a slight decrease during treatment up to rapid sand filtration. Second, the ozone unit process is effective for the removal of all tastes and odors except astringent and plastic. Also, ozonation produces a fruity odor (GAC filtration was only assessed as an add-on to ozonation and not as an individual process to compare to ozonation).

Correlation of Flavor Profile Analysis and Chemical Analysis

Little is known about the nature of the undesirable compounds that cause taste and odor in drinking water. Additional knowledge in this area would allow utilities to select the most appropriate treatment techniques to improve the organoleptic quality of their drinking water and to alleviate taste and odor problems at the time of an episode. The following is an evaluation of the relationships between sensory and chemical analysis before and after ozonation at the Morsang Plant.

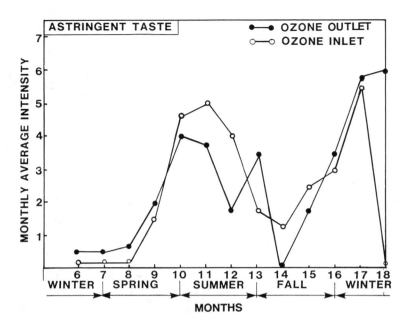

FIGURE 4. (Reprinted from *J. AWWA*, Vol. 88, No. 10 [October 1988]. By permission of American Water Works Association.)

Figure 7 shows the scheme used to study the correlation of sensory and chemical analysis. Organoleptic quality was evaluated using a flavor profile analysis. In this study the most frequently mentioned descriptives were earthy, muddy, musty, fruity, and plastic (see Tables 1 and 2, and Figures 2 to 6).

Chemical analysis initially involved closed loop stripping analysis (CLSA) and was later supplemented by simultaneous distillation extraction (SDE). The CLSA technique used for broad spectrum analysis was similar to that used by Grob and Zurcher.[25] In CLSA, a 1-L aqueous sample is stripped by nitrogen gas for 2 h in a closed loop containing a carbon trap which collects the volatile compounds purged from the sample. The water bath temperature was 45°C and the carbon trap was maintained at 55°C. The carbon disulfide extracts of the carbon filters were analyzed both by gas chromatography/flame ionization detector (GC/FID) and GC/mass spectrometry (GC/MS).

Based on the initial results from the statistical analysis, a second broad spectrum organic technique, SDE,[26,31] was done to complement the data obtained by CLSA. This analytical technique involves a batch extraction of 3 L of water. Water and solvent (methylene chloride) are heated separately, after which they are condensed together. The condensate then separates into two phases and the solvent is recovered for subsequent concentration using a set of small distillation columns (Dufton column). The resulting methylene chloride extracts were

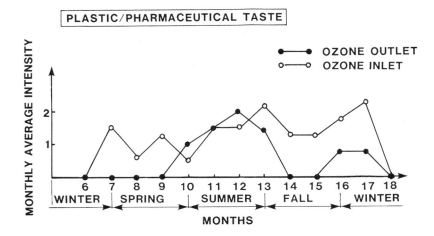

FIGURE 5. (Reprinted from *J. AWWA*, Vol. 88, No. 10 [October 1988]. By permission of American Water Works Association.)

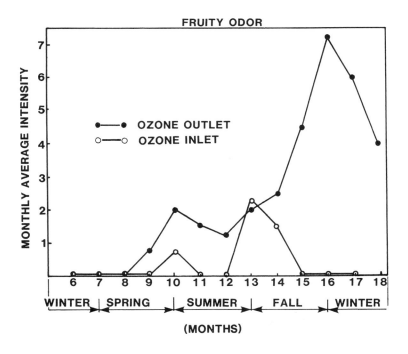

FIGURE 6. (Reprinted from *J. AWWA*, Vol. 88, No. 10 [October 1988]. By permission of American Water Works Association.)

Table 3. Taste and Odor Changes Caused by Ozonation, Morsang Plant (1984)

Response	Line 1 and Line 2 Taste or Odor Intensity
Fruity	Both increase
Musty	Both decrease
Muddy	Both decrease
Earthy	Both decrease
Fishy	Both decrease
Astringent	Taste same
Plastic	Taste same

Reprinted from *J. AWWA*, Vol. 88, No. 10 (October 1988). By permission of American Water Works Association.

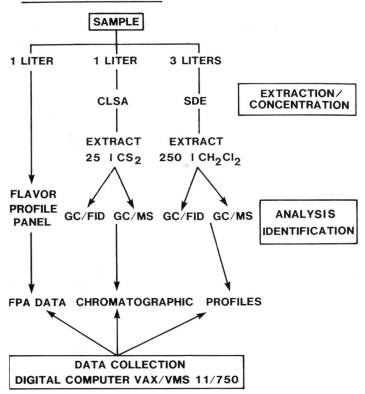

FIGURE 7. (Reprinted from *J. AWWA*, Vol. 88, No. 10 [October 1988]. By permission of American Water Works Association.)

analyzed by GC/FID and GC/MS. By the nature of the technique, SDE is expected to collect compounds of higher molecular weight and higher polarity than CLSA.

Statistical Methods

Statistical analysis was performed by multivariate analysis, as described by Benzecri & Benzecri.[27] The results from the GC/FID analysis were used in the statistical analysis to associate organoleptic properties with "windows" (a specified relative retention time interval) on a chromatogram. The GC/MS results were used to associate identifiable compounds with the windows. For the CLSA chromatograms, the scale of relative retention time was calculated with regard to the retention times of two internal standards (1-chlorooctane and 1-chloredecane). For the SDE chromatograms, the range of relative retention times was narrowed using six internal standards (1-chlorohexane to 1-chlorohexadecane). The chromatograms were divided into 40 windows, each equal to a relative retention time interval of 55 units. Intensities of the organoleptic descriptives and relative areas of chromatographic profiles have been divided in four and three classes, respectively, for the qualitative codification of data. In both cases, responses were grouped into classes such that the ratio of intraclass variance to variance between classes was minimized. The four classes of intensities for FPA data were absence of the descriptives, intensities lower than three, intensities between three and six, and intensities greater than six. The three classes of relative concentrations for CLSA chromatographic data were concentrations less than 80 ng/L, concentrations between 80 and 260 ng/L, and concentrations greater than 260 ng/L.

For the statistical analysis, data from all samples were summarized in a matrix, the elements of which were the number of times a chromatographic response (as classified earlier) was detected concurrent with a particular level of taste or odor.

Correlation between Sensory Analysis and Chromatographic Profiles

All organoleptic descriptives and chromatograms from all CLSA and SDE analyses of the water samples were statistically evaluated using multivariant factor analysis, as described. The statistical analysis associates chromatographic windows with sensory occurrences. GC/MS data are used to identify compounds which are located in specific chromatographic windows.[28]

The purpose of the SDE is to act as a complement to CLSA. A comparison of chromatograms obtained by both these extraction methods is shown in Figure 8. SDE collects a different organic matrix than CLSA, particularly beyond the relative retention time of 1-chlorodecane. The SDE was more effective than

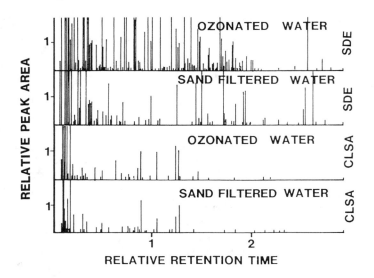

FIGURE 8. Comparison of CLSA and SDE chromatographic profiles at Morsang Line 2. (Reprinted from *J. AWWA*, Vol. 88, No. 10 [October 1988]. By permission of American Water Works Association.)

CLSA in the extraction of polar products which were identified as esters, ketones, aldehydes, phenols, and the extraction of higher molecular weight compounds. Since many of these products are highly odorous compounds, the application of SDE to the CLSA/FPA results is complementary. Figure 8 indicates that many new peaks are produced by ozonation. SDE analysis shows many more peaks and many more new peaks after ozonation.

Tables 4 and 5 summarize the correlations of taste and odor compounds revealed by statistical analysis for the FPA/CLSA and FPA/SDE analyses. Ozonation was found to be responsible for the development of high intensity fruity-type tastes and odors. This descriptive is correlated by factor analysis with an increase in polar product concentrations (e.g., aldehydes, ketones). The descriptives fruity, orange-like, sweet, and fragrant tastes and odors all correlated with aldehydes — straight chain types (CLSA) and benzyl types (SDE). Since a GC retention time window can contain multiple compounds corresponding to an orange-like aroma, it is suspected that the other peaks, e.g., hexane is not the component of primary interest since it does not have anything like an orange-like odor.

The next research step is to test by FPA the individual and complex mixture of aldehydes. Sensory tests on the synergistic effect of the complex mixture that are correlated also should be conducted. Table 5 also shows that the plastic and pharmaceutical tastes that are found are linked with various additives of plastic fabrication. The sensory properties of alkylphenols have been verified by GC/sensory analysis in a recent study.[12,32]

Table 4. Taste and Odor Correlations During Ozone Treatment at Morsang-1

Descriptive	Compounds - CLSA
Orange-like odor	Heptanal
	Hexane
	Methyl ethyl benzene
Fruity odor	Aldehydes C8, C11, C14
Sweet taste	Alkanes C15 to C22
Fragrant odor	Nonanal
Oxidant (T & O)	Heptanal
Sour taste	Trimethylcyclohexane

Reprinted from *J. AWWA*, Vol. 88, No. 10 (October 1988). By permission of American Water Works Association.

Table 5. Taste and Odor Correlations During Ozone Treatment at Morsang-2

Descriptive	Compounds - SDE
Fruity odor	Benzaldehyde
Orange-like odor	Phenyl acetaldehyde
Oxidant (T & O)	Ketones C3 to C10
Plastic odor	4-Methyl-2,6,-Di-*t*-butyl phenol
	1-Ethoxypropane
	2-Methyl-2-ethoxypropane

Reprinted from *J. AWWA*, Vol. 88, No. 10 (October 1988). By permission of American Water Works Association.

Sensory and chemical analysis to develop correlations and point to the direction of the determination of causative agents are being used to help control the ozone treatment process for control of tastes and odors.

CONCLUSION

A combination of chemical and statistical analysis was used to elaborate the source and evolution of tastes and odors in potable water treatment. Ozonation was found to be the first effective unit operation for taste and odor removal in the treatment schemes after chlorination, coagulation, and sand filtration, although this process is responsible for the development of fruity odors with high intensities, in correlation with an increase in aldehyde derivatives concentration.

Ozonation is responsible for the highest decrease of intensity of all descriptors present in the raw water except for an astringent sensation (see Figure 9).

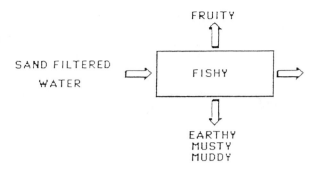
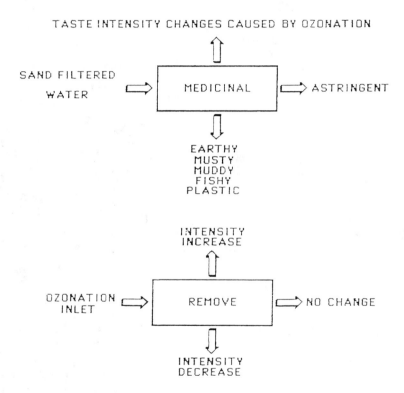

FIGURE 9. (Reprinted from *J. AWWA*, Vol. 88, No. 10 [October 1988]. By permission of American Water Works Association.)

REFERENCES

1. McGuire, M. J. *AWWA J.* June:112 (1986).
2. World Health Organization. Guidelines for Drinking Water Quality. Consultation on Aesthetic and Organoleptic Aspects, ICP/RCE 209 (4), Copenhagen (1981).
3. U.S. Public Health Service. "Public Health Service Drinking Water Standards, 1946," Public Health Report, 61:371 (1946).
4. Public Health Service Drinking Water Standards, PHS Publication No. 956 (1962).
5. Council of the European Communities. "Council Directive of 15 July 1980 Relating to the Quality of Water Intended for Human Consumption," *Off. J. Eur. Communities* 229:11 (1980).
6. APHA. *Standard Methods for Examination of Water and Wastewater,* 16th ed. (New York: American Public Health Association, 1985).
7. Amoore, J. E. "The Chemistry and Physiology of Odor Sensitivity," in *Water Quality Technology Conference Proceeding* (Denver, CO: AWWA, 1984).
8. Cooperative Research Report, AWWARF/Lyonnaise des Eaux, Denver, CO (1987).
9. Arthur, D. Little, Inc. The Flavor Profile Panel. Internal Report, Cambridge, MA (1970).
10. Krasner, S. W., M. J. McGuire, and V. B. Fergusson. "Application of the Flavor Profile Method for Taste and Odor Problems in Drinking Water," presented at the Water Quality Technology Conference, AWWA, Norfolk, VA, December 6, 1983.
11. Bartels, J. et al. "Flavor Profile Analysis: Taste and Odor Control of the Future," *AWWA J.* 78:50 (1986).
12. Anselme, C., K. N'Guyen, A. Bruchet, and J. Mallevialle. "Caractérisation des Produits de Relargage de Canalisations en Polyéthylène Défectueuses: Influence sur les Qualités Organoleptiques des Eaux Transitées," présenté à la Journée CHA de 1 AGHTM, Metz, June 5, 1985.
13. Baker, R. J. et al. presented at the Water Quality Technology Conference Proceeding, AWWA, Portland, OR, 1986.
14. Dore, M. "The Different Mechanisms of the Action of Ozone on Aqueous Organic Pollutants," in *Proceedings of the International Conference — The Role of Ozone in Water and Wastewater Treatment,* R. Perry and A. E. McIntyre, Eds. (London: SP Press, 1985).
15. Hoigne, J., and H. Bader. *Water Res.* 10:377 (1976).
16. Hoigne, J., and H. Bader. 3rd International Ozone Symposium (I.O.A.), Paris, May 4 to 5, 1977.
17. Hoigne, J., and H. Bader. Symposium on Advanced Ozone Technology (I.O.A.), Toronto, November 16 to 18, 1977.
18. Hoigne, J., and H. Bader. *Vom Wasser* 48:283 (1977).
19. Hoigne, J., and H. Bader. *Prog. Water Tech.* 10:657 (1978)
20. Hoigne, J. In: *Ozone Technique and its Practical Application,* Vol. 1, Rice, R. G., and Netzer, A., Eds. (Ann Arbor, MI: Ann Arbor Science, 1984), pp. 341-377.
21. Staehlin, J., and J. Hoigne. *Vom Wasser* 61:337 (1983).

22. Haag, W. R., and J. Hoigne. *Water Res.* 17:1397 (1983).
23. Staehelin, J., and Hoigne, J. *Environ. Sci. Tech.* 19:1206 (1985).
24. Anselme, C., J. Mallevialle, F. Fiessinger, and I. Suffet. Proceedings of the AWWA National Conference, Denver, CO, June 1986.
25. Grob, K., and F. Zurcher. *J. Chromatogr.* 117:285 (1976).
26. Nickerson, G. B., and S. T. Likens, *J. Chromatogr.* 21:1 (1966).
27. Benzecri, J. P., and F. Benzecri. *Pratique de l'Analyse de Données, Vol. 1, Analyse des Correspondances, Exposé Elémentaire.* Bordas. Ed. (Paris: DUNOD, 1980).
28. Anselme, C., K. N'Guyen, J. Mallevialle, and J. P. Bordet. "Influence des Traitements de Désinfection et d'Oxydation sur les Qualités Organoleptiques de l'Eau: Cas de l'Usine de Morsang/Seine," présenté aux 38èmes Journées internationales du CEBEDEAU 85, Bruxelles, June 10 to 12, 1985.
29. Anselme, C., J. Mallevialle, and I. H. Suffet. Proceedings 7th Ozone World Congress, Nihon Toshi Center, Tokyo, Japan, September 9 to 12, 1985, pp. 245-250.
30. Federal Register, "National Secondary Drinking Water Standards," *Fed. Reg.* (July 19, 1979).
31. Nickerson, G. B., and S. T. Likens. *J. Chromatogr.* 11:285 (1976).
32. Persson, E. *Water Res.* 14:1113 (1980).
33. Suffet, I. H. et al. "Taste and Odor in Drinking Water Supplies," Final Report, Year 1 (1984-1985), American Water Works Research Foundation (1986).

CHAPTER 18

The Aqueous Reactions of Specific Organic Compounds with Ozone

N. Graham, C. Corless, G. Reynolds, R. Perry, and J. Haley

ABSTRACT

The reactivity of a range of selected organic compounds, such as alkanes, alkenes, aromatic hydrocarbons, fatty acids, and surfactants, with ozone under idealized conditions has been studied recently. The detectable products of these reactions were identified. Both the reactivity of the compounds and the formation of products are discussed with regard to mechanisms of reaction and previous work on the subject.

INTRODUCTION

The development of sensitive analytical methods for separating and identifying minute quantities of organic substances extracted from water has permitted the accurate determination of many chemicals present in water at or below the nanogram per liter level. In the late 1970s and early 1980s, studies in the United Kingdom[1] showed that many types of organic compounds were present in treated and untreated waters including many pesticides, aromatic and aliphatic hydrocarbons, chlorohydrocarbons, fatty acids, esters, heterocyclic compounds, and surfactants. The identification of such organic chemicals in water has stimulated concern regarding their potential long-term effects on public health, particularly in the light of increasing amounts of re-use of surface water used for drinking water abstraction. Although most of the compounds observed in untreated waters are not considered to pose health hazards to water consumers at the concentrations found, considerable concern exists about the nature of by-products formed by reactions between chlorine (chiefly) and other disinfectants, particularly ozone, and organics in water.

The use of ozone in drinking water treatment is currently widespread in Europe and increasing rapidly in the United Kingdom and the United States. Although the rapid decomposition of ozone to oxygen creates little cause for concern as regards the health effects of ozone or its degradation products, the oxidation products resulting from the reaction of ozone with substances present in raw water are potentially significant. Current information on the nature of these reaction products is very limited.[2,3]

A systematic study is being conducted at Imperial College to investigate the reactivity of a large number of organic substances with ozone. The reactivity of some of these substances is summarized here. Such information is of use in evaluating the public health risks associated with the use of ozone.

EXPERIMENTAL

Methodology

In the study of the reactivity of a specific organic compound with ozone, it is clearly impossible to simulate precisely the reaction conditions that exist in typical water treatment. For example, the presence of other inorganic and organic species in raw waters will directly influence the nature of the reaction mechanism, the extent of compound destruction, and the reaction products formed. As a starting point, in order to observe modes of ozone attack and reaction pathways, investigators typically employ an 'idealized' laboratory methodology. In general, no attempt is made to define and employ 'realistic' concentrations of the specific compound and ozone, and other reactive solutes are not included in the test solution. In our study, the reactivity of specific compounds is being investigated either as single compounds or within a group of similar compounds in a test solution of high-quality water. The initial concentrations of compound (10 µg/L) and ozone (5 mg/L) used were selected as corresponding to the maximum likely to occur in practice. In many cases, ozonation tests were repeated at much higher compound concentrations (e.g., 200 µg/L), which depended on compound solubility, in order to facilitate by-product identification.

Ozonation Procedure

Standard aqueous ozone solutions were prepared by bubbling air containing ozone through high-quality water contained in a recirculating bubble column contactor (Figure 1). Ozone concentrations in aqueous solution were determined by the method of Shechter,[4] and aqueous ozone solutions of approximately 5 mg/L and of pH 4 to 5 typically were produced after 20 min of ozonation. Ozonized water (1 L) was transferred to a reaction flask which was subsequently suspended in a darkened recirculating water bath maintained at 10°C. An appropriate quantity of the organic compound under test together with an internal standard was added in methanol solution to achieve a concentration of 10 μg/L. Procedural blank and unozonized sample solutions were maintained alongside ozonated substrate samples. After 30 min, the reaction was terminated and extraction procedures commenced.

Extraction Procedures

At the end of the ozonation period, the aqueous solution was either extracted with pentane (alkanes, alkenes, and permethrin) or after acidification to pH 2 with a 20% solution of phosphoric acid (10 mL) with diethyl ether (aromatic hydrocarbons, fatty acids, and quaternary ammonium surfactant). When diethyl ether was employed as the extracting solvent, an equal volume of pentane was added to the extract (not in the case of surfactant) prior to drying. Diethyl ether extraction was used instead of pentane because it resulted in improved yields of polar substances from ozonated water. In some cases the extracts were subsequently methylated using diazomethane.

In addition (in the case of pyrene), at the end of the ozonation reaction, dissolved ozone was removed from the sample and control solutions by nitrogen stripping and the sample freeze dried and then methylated using diazomethane. This technique was applied specifically to extract and concentrate short-chain polar aliphatic reaction products.

For the surfactants (except quaternary ammonium surfactant) the ozonized solution was shaken with 100 g dissolved NaCl and extracted with ethyl acetate and dried.

Instrumentation

Gas chromatographic (GC) analyses were performed using a Carlo Erba Strumentazione Fractovap 4200 fitted with a flame ionization detector (FID) on column injector OCI-3 and fused silica capillary columns. Alternatively, a Varian 3700 gas chromatograph fitted with an FID on column injector OCI-3 and fused silica capillary column was employed.

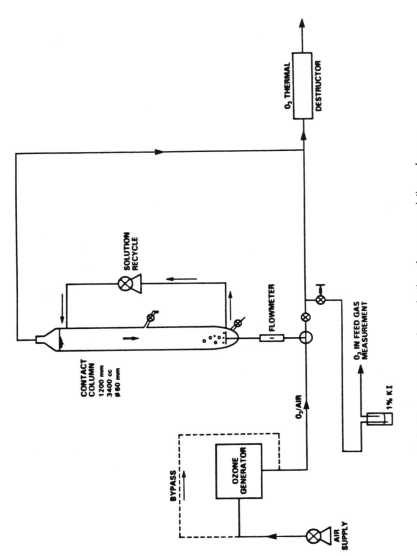

FIGURE 1. Apparatus for production of aqueous solutions of ozone.

Gas chromatography/mass spectrometry (GC/MS) was performed using a Carlo Erba Strumentazione Fractovap 4200 connected to a Jeol JMS-D300 double-focusing mass spectrometer in electron impact mode. Data acquisition and processing was carried out on a Jeol MS DK 400 Disk system. Alternatively, a Hewlett-Packard 5710A GC directly coupled to a VG 70-70E double-focusing mass spectrometer operated in both electron impact and chemical ionization modes was used. Data acquisition and processing was provided by a Super-Incos data system. In both cases mass calibration was achieved using perfluorokerosine.

Fast atom bombardment (FAB) mass spectrometry was carried out on a VG ZAB-1F double-focusing mass spectrometer with data acquisition and processing performed by a Super-Incos data system. A commercial FAB source was used with a saddle-field atom gun operated at 8KV with xenon gas. Mass calibration was achieved using mixtures of rubidium, caesium, and sodium iodides.

High-performance liquid chromatography (HPLC) was carried out using a Waters 6000A HPLC system equipped with an automated gradient elution controller. Detection was by UV absorbance (at 290 nm) and fluorescence spectrometry (excitation wavelength = 230 nm; emission wavelength = 460 nm). The reverse-phase cartridge was operated in a radial compression module with a solvent gradient curve from 35% acetonitrile (65% water) to 92% acetonitrile (8% water) over 40 min at a flow rate of 2 mL min^{-1}.

RESULTS

Alkanes

A standard solution of alkanes was prepared containing the following compounds: undecane, dodecane, tridecane, tetradecane, pentadecane, hexadecane, heptadecane, pristane, octadecane, nondecane, and squalane. Under the conditions employed (10 μg/L of each component, 5 mg/L ozone for 30 min at 10°C), no destruction of any alkanes occurred.

Alkenes

A standard solution of alkenes was prepared containing the following compounds: 3,3,5-trimethylhex-1-ene, non-1-ene, non-4-ene, dec-1-ene, undec-1-ene, dodec-1-ene, tridec-1-ene, tetradec-1-ene, hexadec-1-ene, octadec-1-ene, eicos-1-ene, docos-1-ene, and squalene (2,6,10,15,19,23-hexamethyl-2,6,10,14,18,22-tetracosahexaene). Treatment of the alkene standard solution with ozone under the conditions described (10 μg/L of each alkene) resulted in substantial destruction of the compounds and the formation of a homologous series of products. By comparison of the mass spectra and gas chromatographic

Table 1. Destruction (%) of Aromatic Hydrocarbons by Ozone (5 mg/L)

Compound (10 μg/L)	% Destruction
Naphthalene	49
Phenanthrene	77
Fluoranthene	75
Pyrene	85

retention times of several of these products with those of aldehyde standards, it was established that the products were aldehydes (C_9–C_{14}).

Aromatic Hydrocarbons

A standard solution of aromatic hydrocarbons was prepared containing the following compounds: toluene, *ortho*-, *meta*-, and *para*-xylenes, naphthalene, fluorene, phenanthrene, fluoranthene, and pyrene. After ozonation several of the compounds, namely naphthalene, phenanthrene, fluoranthene, and pyrene were found to have been destroyed to some extent (see Table 1). In addition fluoren-9-one was tentatively identified in the reaction mixture extract obtained after ozonation, thus indicating that fluorene had been consumed to some extent.

In order to further facilitate the identification of products, the reactive aromatic hydrocarbons (viz. naphthalene, fluorene, phenanthrene, fluoranthene, and pyrene) were ozonated individually at a much higher compound concentration (1, 1, 1, 0.2, and 0.2 mg/L, respectively). From this, phthalic acid has been tentatively identified as a product of naphthalene ozonation by comparison with the mass spectrum of a library standard held on the data system. Fluoren-9-one has been confirmed as a product of the ozonation of fluorene by comparison of both gas chromatographic retention time and mass spectrum with those of a standard. To date, no products of ozonation of phenanthrene, fluoranthene, and pyrene have been identified. Further study of pyrene[5] has shown the absence of any polyhydroxylated aromatic compounds or polymeric species in the pyrene reaction solution.

Fatty Acids

A standard solution of saturated and unsaturated fatty acids was prepared containing the following compounds: octanoic, decanoic, dodecanoic, tetradecanoic, hexadec-9-enoic, hexadecanoic, heptadecanoic, octadec-9,12-dienoic, and octadecanoic acids. After ozonation little or no destruction of the saturated fatty acids was evident, but near complete destruction (at least 95%) of the unsaturated fatty acids, hexadec-9-enoic, and octadec-9,12-dienoic acids was observed. In order to identify the products of the ozonation of unsaturated acids, ozonation was repeated using a higher initial compound concentration (viz. 500 μg/L) and dodecanoic acid as an internal standard.

After ozonation and extraction, a portion of the product was methylated and analyzed by gas chromatography. Among the compounds found in ozonated solutions of the unsaturated fatty acids but not in control solutions, were hexanal and heptanal together with their methylated derivatives heptan-2-one and octan-2-one. In addition smaller amounts of the methyl esters of hexanoic and heptanoic acids and two other products, namely 9-oxononanoic acid methyl ester and 9-oxodecanoic acid methyl ester, were tentatively identified. Support for the identification of 9-oxononanoic acid methyl ester was provided by negative ion FAB mass spectral analysis of the unmethylated extracts obtained with and without ozonation. FAB mass spectra obtained after ozonation showed ions corresponding to 9-oxononanoic acid and nonanedioic acid. Accurate mass measurements using peak matching techniques demonstrated that the ions were consistent with empirical formulae of $C_9H_{15}O_3$ (9-oxononanoic acid) and $C_9H_{15}O_4$ (nonanedioic acid). From these data, it is evident that unsaturated fatty acids undergo ozonolysis followed by some degree of oxidation during treatment with ozone under aqueous conditions.

Surfactants

The reaction of a range of surfactants with ozone was studied. Two cationic (viz. quaternary ammonium surfactant and capyl pyridinium chloride), one anionic (viz. 4-dodecyl phenyl sulfonate) and one nonionic (viz. nonyl phenyl heptakis ethoxylate) were included.

Extracts produced following the ozonation of 4-dodecyl phenyl sulfonate (compound concentration 240 μg/L) were analyzed by FAB negative ion mass spectrometry. The presence of 4-dodecyl phenyl sulfonate (DPS) anion in both ozonated and unozonated sample spectra and the absence of reaction product anions indicated no reaction between DPS and ozone.

Extracts produced following the ozonation of nonyl phenyl heptakis ethoxylate (compound concentration 240 μg/L) were analyzed by FAB positive ion mass spectrometry. The absence of the appropriate peak in the FAB positive ion mass spectrum suggested the destruction of this surfactant, although no reaction product ion peaks were apparent. However, after analysis by GC/MS, a series of as yet unidentified products were observed.

The ozonation of the cationic surfactants was carried out at a compound concentration of 500 μg/L and extracts were analyzed by FAB mass spectrometry. In the case of capyl pyridinium chloride (CPC), the presence in both ozonated and unozonated sample FAB positive spectra of the ion corresponding to the CPC cation, and the lack of reaction product cations, suggested no reaction between CPC and ozone.

The quaternary ammonium surfactant was obtained as a commercial sample and consisted of dimethyl dialkyl and trimethyl alkyl ammonium chlorides containing a mixture of unsaturated and saturated side chains. Using FAB mass spectrometry, ions corresponding to the unsaturated and saturated components were observed in the unozonated sample extract, but following ozonation, ions corresponding to the unsaturated components largely disappeared. The appearance of a number of ion peaks indicated the existence of reaction products. Accurate mass measurements of some of these products suggested that their compositions were consistent with quaternary ammonium carboxylic acids. GC/FIC and GC/MS analysis of extracts showed that a number of reaction products were present in the ozonated sample extract. Nonanal, together with 2-decanone and methyl nonanoate, the methylated derivatives of nonanal and nonanoic acid, respectively, were identified by GC/MS.

Permethrin

Ozonation of the pesticide, permethrin, was carried out with a compound concentration of 10 μg/L. Both *cis*- and *trans*-permethrin were destroyed by >90%, but no ozonation products were found.

DISCUSSION

Mechanisms proposed for the reaction of ozone with organic compounds have been divided into two categories[6] involving indirect (or free radical) reactions and direct reactions. The indirect mechanism arises from the decomposition of ozone in water favored by any of the following conditions: increasing hydroxyl ion concentration, i.e., pH increase; the presence of hydrogen peroxide; and photolysis by UV radiation.[6] The neutral/slightly acidic conditions used in this study should result in reactions principally via a direct mechanism with ozone (as proposed by Decoret et al.[7]) which are specific in nature and relatively slow.

Direct reactions with ozone are a consequence of the resonance structure of the ozone molecule which forms an electric dipole. It is however insufficiently reactive to attack saturated aliphatic hydrocarbons, as is demonstrated by the lack of consumption of alkanes and saturated fatty acids observed in the present study. In contrast, the direct reaction of ozone is well illustrated by its 1,3-dipolar cycloaddition across unsaturated double bonds, such as in alkenes and unsaturated fatty acids, in which an ozonide (1,2,3-trioxolane) is first produced. In aqueous solution at 10°C, the ozonide decomposes to form a hydroxyhydroperoxide[8] which can undergo further decomposition to form hydrogen peroxide and carbonyl compounds (aldehydes and ketones). Aldehydes so formed can be oxidized either by ozone or the hydrogen peroxide produced to form carboxylic acids and peracids. In this study the action of ozone on alkenes, unsaturated fatty acids and the unsaturated components of the quaternary ammonium surfactant, appears to follow this course with aldehydes or aldehydoacids being produced in relatively large yields together with much smaller quantities of carboxylic acids and diacids. As examples, the proposed reaction pathways for the unsaturated fatty acid, 9-hexadecenoic acid, and the unsaturated components of the quaternary ammonium surfactant are shown in Figure 2 and Figure 3, respectively. These have been discussed in detail elsewhere.[9,10]

Various authors have described the reaction between ozone and aromatic compounds in terms of a two- or three-stage process (Figure 4).[7,11] The first stage involves the formation of polyhydroxy aromatic species and loss of some or all of the aromaticity via electrophilic attack. Subsequent stages involve the ozonation of the nonaromatic products of the first stage. In the present study the tentative identification of phthalic acid from the ozonation of naphthalene and fluoren-9-one from fluorene are consistent with observations reported in the literature.[11-13] In the case of pyrene, although no products were observed, there appeared to be a complete loss of aromaticity.

The absence of reaction observed for capyl pyridinium chloride and 4-dodecyl phenyl sulfonate may be attributable to the presence of the pyridinium ion and the sulfonate group, respectively, deactivating the aromatic ring system and making it less susceptible to ozone attack. In contrast, the reactivity of the nonionic surfactant, nonyl phenyl heptakis ethoxylate, may be due to ring activation, and thus electrophilic attack, arising from the presence of the alkyl side chain (weakly activating) and the ether group (moderately activating).[14] A previous study by Narkis et al.[15] has suggested that the ozonation mechanism is mainly cleavage of the polyethoxylate chain adjacent to the ether groups and to a lesser extent oxidation of the aromatic ring system. However, rupture of the ether linkage (C–O–C) is unlikely since severe conditions are required for such a reaction.[14]

Considerable reactivity of permethrin towards ozone has been demonstrated, but reaction products have not been identified. Three possible sites of attack by ozone are evident from the structure of permethrin, namely electrophilic attack of the aromatic ring system, 1,3-dipolar cycloaddition across the carbon-carbon double bond, and reaction at the cyclopropane ring because of the inherent angle strain of the ring system.

CONCLUSIONS

This study has shown that substantial reaction takes place between ozone and a number of organic compounds containing aromatic or alkene functional groups. Such compounds are commonly occurring contaminants of water and reaction leads to the formation of new compounds whose significance is as yet unknown. The isolation and identification of these reaction products, particularly from aromatic compounds, has proved difficult.

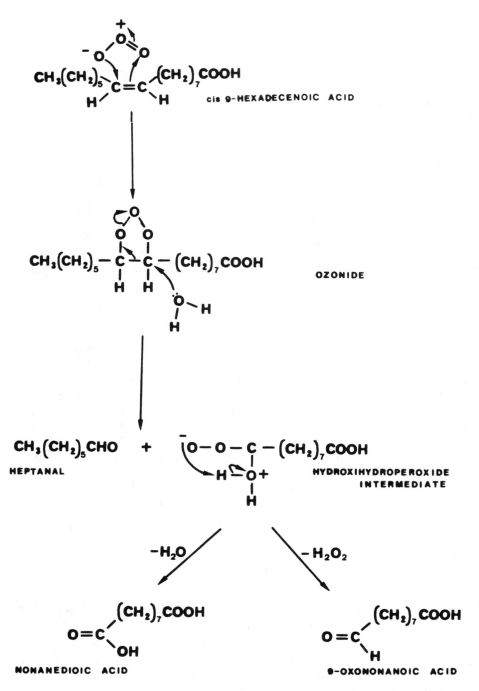

FIGURE 2. Proposed pathway for the reaction between ozone and 9-hexadecenoic acid.

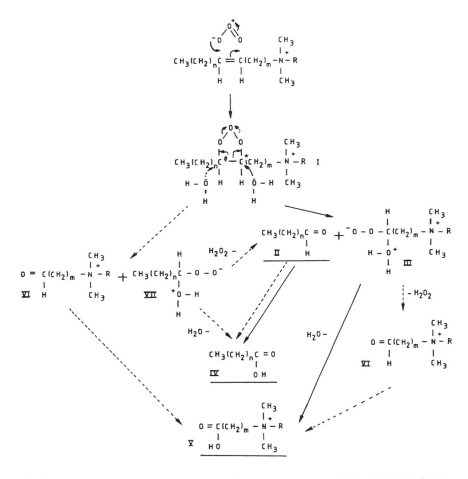

FIGURE 3. Proposed pathway for the reaction between ozone and the unsaturated components of the quaternary ammonium surfactant. Alternative reaction pathways are represented by broken lines. Compounds of the general formulae shown underlined were identified in this study.

ACKNOWLEDGMENT

The work described in this paper was funded by the U.K. Department of the Environment.

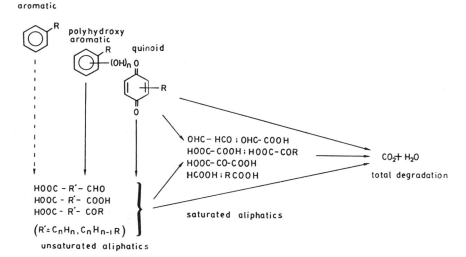

FIGURE 4. General scheme for ozonation of aromatics.[7]

REFERENCES

1. Fielding, M., Gibson, T. M., James, H. A., McLoughlin, K., and Steel, C. P. "Organic Micropollutants in Drinking Water," Water Research Centre Technical Report TR159 (1981).
2. Fielding, M., Haley, J., Watts, C. D., Corless, C., Graham, N. J. D., and Perry, R. "The Effect of Chlorine and Ozone on Organic Compounds in Water — a Literature Review," *Water Res. Cent.* PRD 1217-M, U.K. (1987).
3. Reynolds, G., Graham, N. J. D., Perry, R., and Rice, R. "Aqueous Ozonation of Pesticides: a Review," *Ozone Sci. Eng.* 11(4):339–382 (1989).
4. Shechter, H. "Spectrophotometric Method for Determination of Ozone in Aqueous Solutions," *Water Res.* 7:729-739 (1973).
5. Corless, C., Reynolds, G., Graham, N., and Perry, R. "Ozonation of Pyrene in Aqueous Solution," *Water Res.* 24:1119-1124 (1990).
6. Doré, M. "The Different Mechanisms of the Action of Ozone on Aqueous Organic Micropollutants," in *Proceedings of the International Conference: the Role of Ozone in Water and Wastewater Treatment,* R. Perry and A. McIntyre, Eds. (London: 1985), pp. 321-336.
7. Decoret, C., Roger, J., Legube, B., and Doré, M. "Experimental and Theoretical Studies of the Mechanism of the Initial Attack of Ozone on Some Aromatics in Aqueous Medium," *Environ. Tech. Lett.* 5:207-218 (1984).
8. Bailey, P. S. and Philips, S. "Olefinic Compounds," in *Organic Chemistry Monographs* (New York: Academic Press, 1978), p. 272.
9. Reynolds, G., Corless, C., Graham, N., Perry, R., Gibson, T. M., and Haley, J. "Aqueous Ozonation of Fatty Acids," *Ozone Sci. Eng.* 11:143-154 (1989).

10. Corless, C., Reynolds, G., Graham, N., Perry, R., Gibson, T. M., and Haley, J. "Aqueous Ozonation of a Quaternary Ammonium Surfactant," *Water Res.* 23:1367-1372 (1989).
11. Legube, B., Guyon, S., Sugimitsu, H., and Doré, M. "Ozonation of some Aromatic Compounds in Aqueous Solution: Styrene, Benzaldehyde, Naphthalene, Diethyl Phthalate, Ethyl, and Chlorobenzenes," *Ozone Sci. Eng.* 5:151-170 (1983).
12. Legube, B., Guyon, S., Sugimitsu, H., and Doré, M. "Ozonation of Naphthalene in Aqueous Solution I and II," *Water Res.* 20:197-214 (1986).
13. Helleur, R., Malaiyandi, M., Benoit, F. M., and Bonedek, A. "Ozonation of Fluorene and 9-Fluorenone," *Ozone Sci. Eng.* 1:249-261 (1979).
14. McMurry, J. *Organic Chemistry,* (CA: Brooks/Cole Publishing Company, 1984).
15. Narkis, N., Ben-David, B., and Schneider-Rotel, M. "Ozonation of Non-Ionic Surfactants in Aqueous Solutions," *Water Sci. Tech.* 17:1069-1080 (1985).

CHAPTER **19**

New Advances in Oxidation Processes: The Use of the Ozone/Hydrogen Peroxide Combination for Micropollutant Removal in Drinking Water

J. P. Duguet, A. Bruchet, and J. Mallevialle

INTRODUCTION

Ozone has been used for many years in water treatment for different purposes: tastes and odors control, iron and manganese removal, disinfection, micropollutants removal, etc. Recent progress in ozone chemistry has allowed us to understand the oxidation mechanisms.[1] Two different pathways are involved during the ozonation: a direct oxidation by the ozone molecule which is very selective and indirect radical-type reactions by mainly hydroxyl radicals resulting from the ozone decay. Oxidation by hydroxyl radicals is nonselective and very efficient toward organics removal. Many processes are able to generate hydroxyl radicals among them: ozone at high pH, ozone combined to hydrogen peroxide,[2,3] photolysis of ozone,[4] or hydrogen peroxide.[5] Due to the easy application of the ozone/hydrogen peroxide combination on existing water treatment plants equipped with ozonation steps, a lot of research was done in order to test the feasibility of such combinations to increase the water quality. In this paper, the results of three recent laboratory or pilot studies are reported.

THE OZONE-HYDROGEN PEROXIDE COMBINATION

Hydrogen peroxide reacts with ozone, resulting in an increase of ozone decay. Kinetic experiments performed by Staehelin[6] have shown that H_2O_2 reacts very slowly with ozone, suggesting that the reactive species is in its ionized form.

$$H_2O_2 + H_2O \Leftrightarrow H_3O^+ + HO_2^- \quad pK\ H_2O_2: 11.6$$

$$O_3 + H_2O_2 \xrightarrow{\text{very slow}} \quad kO_3, H_2O_2 < 10^{-2}\ M^{-1}\ s^{-1}$$

$$O_3 + HO_2^- \xrightarrow{\text{fast}} \quad kO_3, HO_2^- = 5.5 \times 10^6\ M^{-1}\ s^{-1}$$

The rate of ozone decay increases 10-fold per pH unit. The main reaction mechanism is done below:

$$O_3 + HO_2^- \rightarrow OH^\cdot + {}^\cdot O_2^- + O_2$$

$$O_3 + {}^\cdot O_2^- \rightarrow {}^\cdot O_3^- + O_2$$

$${}^\cdot O_3^- + H_2O \rightarrow OH^\cdot + OH^- + O_2$$

Different species such as ozonide ion-radical ${}^\cdot O_3^-$ are formed and lead to the hydroxyl radical production. Mineral scavengers (bicarbonates) present in treated water can drastically reduce the efficiency of pollutant oxidation proportionally to their concentrations.

$$OH^\cdot + HCO_3^- \rightarrow OH^- + HCO_3^\cdot$$

$$OH^\cdot + CO_3^{2-} \rightarrow OH^- + {}^\cdot CO_3^-$$

$$OH^\cdot + PO_4^{3-} \rightarrow OH^- + {}^\cdot PO_4^{2-}$$

In conclusion, the efficiency of the oxidation by radical will depend on the dissolved ozone and hydrogen peroxide concentrations, the pH, the nature and concentration of pollutant to be removed, and the concentration of scavengers such as bicarbonates. The three examples reported here illustrate these conclusions.

REMOVAL OF AROMATIC COMPOUNDS FROM GROUNDWATER

Nitro and chloro-benzenic compounds, which are widely used in dye industries, have been associated recently with ground water contamination. Because of their potential toxicity and for taste and odor considerations, the feasibility of using the ozone/hydrogen peroxide combination was determined on a pilot plant.

Raw Water Characteristics

The pilot plant is fed by ground water from a well. The pH of the water is about 6.7 and the alkalinity is equal to 225 mg $CaCO_3$/L. During tests, the total concentration of benzenic compounds was in the range of 1400 to 2500 µg/L, with an average value equal to 2100 µg/L. The major pollutant (representing 70% of the pollution) is o-chloronitrobenzene (1000 to 1800 µg/L, average concentration, 1500 µg/L). Some other pollutants, such as chlorobenzene and nitrophenol, are present in the ground water at variable concentrations depending on the sampling period.

During the test periods, threshold odor numbers (TON) in the fed water were above 20.

Pilot Plant

The study is performed on a 450-L/hr pilot. This pilot plant includes two ozone contactors (150 mm in diameter, 4 m in height), an ozone generator (Degrémont, France) which produces up to 10 g/hr, a dosage system for hydrogen peroxide which permits the injection of a diluted solution in the input of the two columns, and two filters running in parallel (sand and granular activated carbon) after the oxidation step to reduce biodegradable oxidation by-products.

Results

Two types of tests are performed to verify the efficiency of ozone alone and ozone combined with hydrogen peroxide. Geosmin and 2-MIB which are tertiary alcohols are nonreactive towards molecular ozone, but can be removed by hydroxyl radical formed during the ozone decomposition in water. So laboratory studies were performed in order to test the feasibility of an ozone/hydrogen peroxide combination for increasing the removal efficiency.

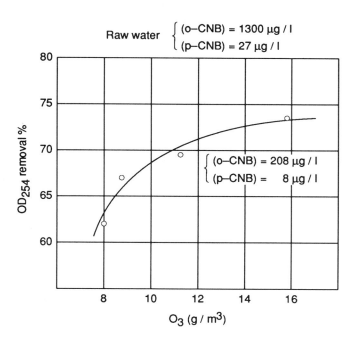

FIGURE 1. Detoxification of ground water. Influence of the ozone dosage (contact time: 30 min).

Ozonation

The results obtained by ozonation alone are summarized in Figures 1 and 2. The tests show that for a contact time of 30 min, the decrease in optical density at 254 nm (which corresponds to the decrease in aromatic compound concentrations) is a function of the increasing applied ozone dose. At an ozone dose of 16 g/m^3, the removal of chloronitrobenzene reaches 85%. With a constant ozone dose, the reduction of the optical density at 254 nm increases when the contact time is longer, up to 30 min. The application of 16 g O$_3$/m^3 with a 40-min contact time allows an 88% reduction of chloronitrobenzene corresponding to a residual concentration of 150 μg(o-chloronitrobenzene)/L in the ozonated water. Oxidation by ozone alone, even for large doses of ozone and long contact times, does not allow a sufficient reduction of the chloronitrobenzene concentration, which remains higher than the desired amount of 20 to 30 μg/L in the ozonated water.

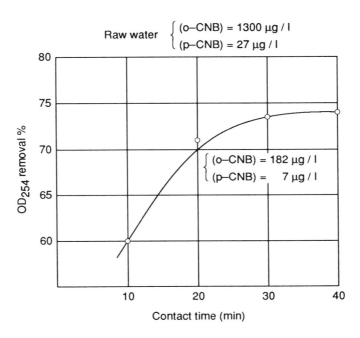

FIGURE 2. Detoxification of ground water. Influence of contact time (ozone dose: 16 g/m³).

Combination of Ozone with Hydrogen Peroxide

The O_3/H_2O_2 combination is performed in two columns placed in series. Operating parameters are the same in each column. Previous experiments reported in Figure 3 show that the best o-chloronitrobenzene removal is obtained for a hydrogen peroxide/ozone mass ratio of 0.4. The application of ozone and hydrogen peroxide dosages of 8 and 3 g/m³, respectively, during a 20-min contact time permitted a 99% elimination of benzenic compounds (Figure 4). An increase in the contact time to 30 min does not improve removal efficiency.

Efficiency of the Complete Treatment Line

It is well-known that during the oxidation process biodegradable compounds are formed which can induce bacterial regrowth in the distribution network. To avoid this, the oxidation process must be followed by a biological treatment such as sand or GAC filtration. The latter technique has a specific advantage in that it removes micropollutants by adsorption. For this study, GAC and sand columns with 15-min hydraulic residence times are used. All of the treatment lines run

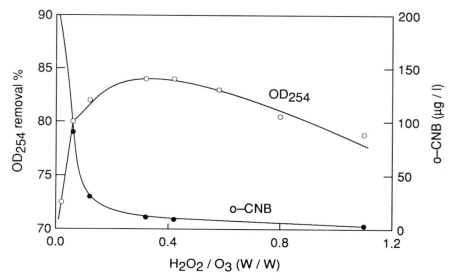

FIGURE 3. Detoxification of a ground water influence of H_2O_2 dosage on the oxidation efficiency (ozone dosage: 12 g/m^3; contact time: 20 min).

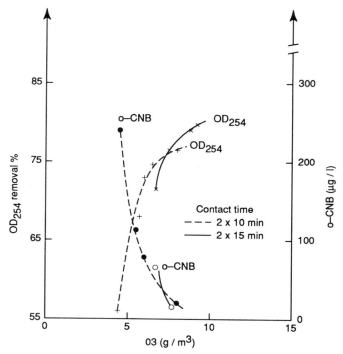

FIGURE 4. Detoxification of a ground water, removal of DO_{254} and o-CNB by a double O_3/H_2O_2 injection. Influence of the ozone dose and contact time (H_2O_2/O_3: 0.4 w/w; [o-CNB]$_o$ = 1800 µg/L.

Table 1. Detoxification of a Ground Water — Evolution of TOC, Benzenic Compounds, o-Chloronitrobenzene and Threshold Odor Number at Different Treatment Steps

Running Time (Days)	0	4	11	19	25	32	39
Raw water							
TOC (mg/L)	1.9		1.9	1.95	1.75	1.9	
Benzenic compounds (µg/L)	2023		1397	2249	2031	1750	2121
o-Chloronitrobenzene (µg/L)	1690		970	1830	1490	1320	1402
TON > 20							
Oxidized water							
TOC (mg/L)	1.7	1.5	1.75	1.6	1.3	1.45	
Benzenic compounds (µg/L)	52	6	16	21	18	19	23
o-Chloronitrobenzene (µg/L)	41	4	13	14	14	14	14
TON = 3 to 5							
Sand-filtered water							
TOC (mg/L)	1.4	1.55	1.45	1.55	1.15	1	
Benzenic compounds (µg/L)	17	6	11	35	17	11	14
o-Chloronitrobenzene (µg/L)	15	3	8	28	11	8	9
TON = 2 to 3							
GAC-filtered water							
TOC (mg/L)		0.9	0.40	0.40	0.35	0.45	
Benzenic compounds (µg/L)		<1	<1	<1	<1	<1	<1
o-Chloronitrobenzene (µg/L)		<1	<1	<1	<1	<1	<1
TON = 1							

for 3 months. Samples are collected weekly at different treatment steps. The mean concentrations of benzenic compounds and o-chloronitrobenzene in the water after the oxidation step are 22 and 16 µg/L, respectively, corresponding to a 99% reduction. TOC removal obtained by oxidation is about 0.3 mg/L; this corresponds to the mineralization of 30% of the benzenic compounds (Table 1). Identification of oxidation by-products (e.g., aldehydes, carboxylic acids) will follow at a later stage of the study. After GAC filtration, the treated water has a benzenic compounds concentration lower than the detection limit (1 µg/L). From raw water to oxidized water, the threshold odor number (TON) is reduced from 20 to 3, and there is no taste and odor in the GAC filtered water after 3 months of operating time. Therefore, this treatment line is very efficient and results in a high-quality drinking water.

REMOVAL OF GEOSMIN AND 2-MIB FROM DRINKING WATER

Among the several tastes and odors present in drinking water the most frequently encountered are those which are earthy and musty. Many compounds can be the source of such tastes and odors, among them geosmin and 2-methylisoborneal (2-MIB), which are produced by living algae. Conflicting results concerning their removal by ozone have been reported in the literature. Geosmin and 2-MIB, which are tertiary alcohols, are nonreactive towards molecular ozone, but can be removed by hydroxyl radical formed during the ozone decomposition in water. So laboratory studies were performed in order to test the feasibility of an ozone/hydrogen peroxide combination for increasing the removal efficiency.

Test Solutions

Standard solutions of geosmin and 2-MIB were prepared with pure water. Geosmin and 2-MIB (98% grade Wako Chemicals, West Germany) were initially dissolved in pure acetone (10 μg/L of acetone). Pure water was then spiked with these two solutions and stirred during half an hour in order to prepare test solutions at 300 to 500 ng/L compound concentration. The pH of some of the test solutions was then adjusted using 0.1 or 1 N NaOH.

Experimental Setup

Oxidation tests were carried out in a 10-L semibatch reactor of 150 mm diameter and 700 mm high. Ozone was produced by a 10-g/hr ozonator supplied with pure air (Degrémont). The ozone concentration in the gas was around 2 mg O_3/NL (CO_3) and the gas flow rate 40 NL/hr (G). The reaction of ozone with organics is generally first order with respect to the dissolved ozone and compound concentrations. In order to establish the oxidation kinetics, tests at acidic and neutral pH were performed at a constant dissolved ozone concentration (0.2 to 0.3 mg O_3/L), which was reached in the first minutes of ozonation. 1 L of ozonated solution was sampled at regular intervals of time, the ozonation reaction was then stopped by adding some drops of sodium thiosulfate solution. Remaining geosmin and 2-MIB were analyzed using the close loop stripping technique coupled to gas chromatography/mass spectrometry.

Results

Effect of pH on Geosmin and 2-MIB Removal by Ozone

Figure 5 shows that geosmin is removed by ozonation alone. At pH 5.6 and 7.5 the concentration of geosmin is reduced from 350 to 500 ng/L to around 100 ng/L in 30 min. At pH 11 the reaction rate is five- to sixfold faster than at pH 5.6 and 7.5, so the geosmin is reduced to less than 10 ng/L with 20 min ozonation time. Similar results are obtained with 2-MIB. As mentioned previously, geosmin and 2-MIB are not oxidized by the ozone molecule. So the removal observed may be due to the formation of hydroxyl radicals during ozone

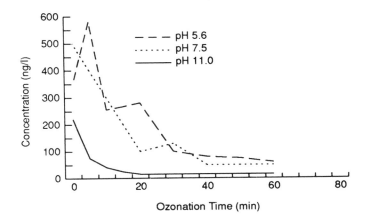

FIGURE 5. Influence of pH on the geosmin removal by ozonation in pure water (CO_3: 2 mg/L — G: 40 Nl/hr).

decomposition by hydroxyl ions (OH^-) or promotors present at very low concentrations in the "pure" water. The higher removal rate observed at pH 11 can be explained by the higher rate of ozone decomposition at basic pH, producing higher hydroxyl radical flux and perhaps by the better reactivity of ionized form of the alcohol group on geosmin and 2-MIB.

REMOVAL OF GEOSMIN AND 2-MIB BY THE OZONE/HYDROGEN PEROXIDE COMBINATION

Pure water at pH 7.5 was used to perform ozonation tests with hydrogen peroxide. In the presence of hydrogen peroxide, the removal rate of geosmin increases drastically so it is possible to reach a geosmin concentration below 10 ng/L in less than 5 min of ozonation time (Figure 6). The increase of hydrogen peroxide concentration has no influence on the removal rate. In these experiments the hydrogen peroxide concentration has not been optimized. A previous study has shown that the optimal H_2O_2/O_3 mass ratio is about 0.5. Thus, in these experiments, 1 mg O_3 per L of water is applied and the expected optimal hydrogen peroxide concentration will be 0.5 mg/L. Similar results were obtained for the 2-MIB removal.

Influence of Bicarbonate Ions on 2-MIB Removal

The use of the ozone/hydrogen peroxide combination for the removal of compounds in water can be limited by the hydroxyl radical scavenging effect of

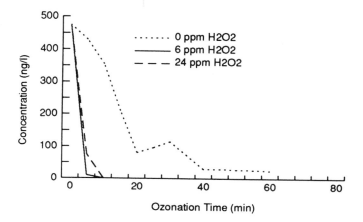

FIGURE 6. Influence of the hydrogen peroxide concentration on the elimination of geosmin by ozonation in pure water at pH 7.5 (Co_3: 2 mg/Nl — G: 40 Nl/hr).

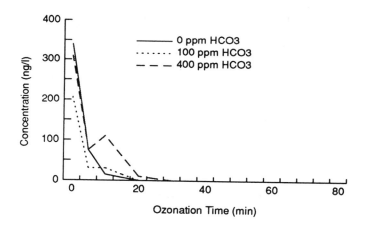

FIGURE 7. Influence of bicarbonate concentration on the removal of 2 MIB by O_3/H_2O_2 in pure water (Co_3: 2 mg/Nl — G: 40 Nl/hr).

bicarbonates present in water. In order to test the influence of such ions on the 2-MIB and geosmin removal efficiency, experiments were conducted with the O_3/H_2O_2 combination in pure water at pH 7.5 with different bicarbonate concentrations (0 to 400 mg/L). Under these conditions for geosmin and 2-MIB, bicarbonates do not have a significant effect on the removal efficiency (Figure 7). Similar results have been found by comparing the removal rate of 2-MIB by ozonation performed without hydrogen peroxide in pure water at pH 7.5 and this holds true in tap water at the same pH, but with an alkalinity equal to 250 mg/L as $CaCO_3$.

REMOVAL OF TRICHLORETHYLENE AND TETRACHLORETHYLENE FROM GROUND WATER

Ground water is very often polluted by chlorinated solvents such as trichlorethylene (TCE) and tetrachlorethylene (PCE). New regulations concerning the maximum contaminant levels boost the application of technologies for removing such compounds. In the case of chlorinated solvents, two technologies are available: air stripping and/or granular activated carbon (GAC) filtration. Ozonation alone doesn't permit the oxidation of chlorinated solvents. The oxidation by hydroxyl radical generated by, for example, the ozone/hydrogen peroxide coupling is, however, a promising method to be evaluated.

Raw Water Characteristics and Experimental Setup

Tests were performed with ground water. The alkalinity and the pH of such water are 262 mg as $CaCO_3$/L and 7.1, respectively. During tests, the TCE and PCE concentration ranges were, respectively, 65 to 85 and 10 to 20 µg/L.

In this study a 15-L reactor was used in continuous flow, hydrogen peroxide was added before the reactor inlet at a rate allowing a constant H_2O_2/O_3 mass ratio equal to 0.4. The influence of contact time, gas flow rate, and ozone concentration were evaluated.

Results

Tests performed with this reactor without ozone have shown that no significant quantities of chlorinated solvents are removed by stripping. Ozonation alone has not produced the removal of more than 30% of chlorinated solvents even for an applied ozone dosage up to 8 g/m^3 and 10 min contact time (Figures 8 and 9).

The ozone/hydrogen peroxide coupling achieves, under these test conditions, a 98% TCE removal superior and PCE removal so the TCE and PCE concentrations in the effluent are below 2 µg/L. This high efficiency can be reached using applied ozone doses equal to approximatively 5 mg/L. Particularly in the case of TCE, removal efficiency depends directly on the applied ozone dose up to 5 mg/L. It was not possible to conclude on the contact time influence, but it can be expected that contact time has a small effect because hydroxyl radical O_3/H_2O_2 reactions are relatively fast and the global reaction rate is limited by the ozone transfer.

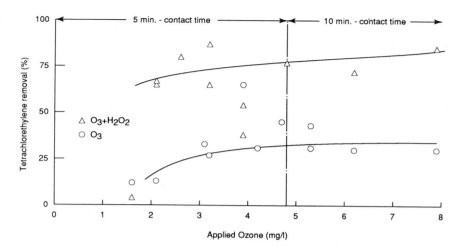

FIGURE 8. Influence of ozone dose on tetrachlorethylene removal.

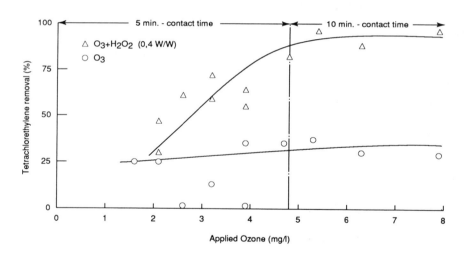

FIGURE 9. Influence of ozone dose on trichlorethylene removal.

CONCLUSIONS

The results of the three studies concerning the removal of organics by the ozone-hydrogen peroxide coupling show clearly that the hydroxyl radical enhances drastically the organic compounds oxidation. The use of this technique allows new regulatory goals to be met. On the other hand, the oxidation rate enhancement permits the reduction of the size of oxidation equipment, thus saving space.

Due to the hydroxyl radical scavenging by the bicarbonate ions, the efficiency of the process and the applied oxidant doses depend largely on the water alkalinity. The ozone-hydrogen peroxide process has a major advantage when the drinking water treatment plant is equipped with an ozone process: the addition of hydrogen peroxide is very easy to perform, allowing a reduction of accidental or chronic pollution.

REFERENCES

1. Hoigné, J. and H. Bader. *Water Res.* 10:377-386 (1976).
2. Duguet, J.-P., E. Brodard, B. Dussert, and J. Mallevialle. *Ozone Sci. Eng.* 7:241-258 (1985).
3. Duguet, J.-P., C. Anselme, P. Mazounie, and J. Mallevialle. "Application of the Ozone-Hydrogen Peroxide Combination for the Removal of Toxic Compounds from Groundwater, in *Organic Micropollutants in the Aquatic Environment,* Commission of European Communities (Dordrecht, The Netherlands: Kluwer Academic Publisher, 1987).
4. Duguet, J.-P., C. Brossard, E. Brodard, and J. Mallevialle. "Development in Ozonation Techniques using Combination of Ozone and Ultraviolet Light," Proceedings 8th IOA Ozone World Congress.
5. Guittonneau, S., J. De Laat, M. Dore, J. P. Duguet, and C. Bonnel. *Environ. Technol. Lett.* 9:1115-1128 (1988).
6. Staehelin, J. and J. Hoigné. *Environ. Sci. Technol.* 16:676-681 (1982).

]

CHAPTER 20

A Fluidized Biofilter Bed Process as a Preliminary Treatment of a Contaminated Raw Water Source

Y. Zheng-Zhong and Z. Hua

ABSTRACT

This experimental study was carried out with the new water treatment technology of fluidized biofilters. The fluidized biofilter is used as a preliminary treatment process for contaminated raw water sources. After a pilot-plant study, further advanced studies on an intermediate-size scale were completed to simulate plant production at a water works.

The result of pilot-plant studies indicates that the fluidized biofilter process effectively removed ammonia-nitrogen from Huang-pu river water. Organic substances and heavy metals in the water were also reduced. Thus, the fluidized biofilter improved the quality of water. During the experimental studies, the techniques of oxygen recharge and fluidization of bed were also evaluated.

INTRODUCTION

The problem of the pollution of surface water sources that are used for drinking water has gained worldwide attention. The protection of drinking water sources and the improvement of water treatment technology have both become methods to remedy the increasing pollution of drinking water sources. The water source of the Huang-pu River of Shanghai has been heavily polluted. The majority of the polluting substances are ammonia-nitrogen (NH_3–N) and other oxygen-consuming organic chemicals.

The content of NH_3–N in water is an index of contamination of the raw water source that is polluted with organic matter and it directly influences the dose of chlorine and the purification of drinking water. The usual process employed for the treatment of polluted raw water at the various waterworks in Shanghai are breakpoint chlorination, or ozonation and activated carbon adsorption. However, the latter method is still in the experimental stage. Normally the dosage of chlorine during breakpoint chlorination is equal to 10 times the ammonia-nitrogen in water. The excessive use of chlorine can cause serious problems, e.g., safety problems during administration of chlorine at the water plant to cancel it. Chlorination can also produce a considerable amount of halogenated compounds in water which are nuisances or potential health problems. This is why the biooxidation technology used for preliminary treatment of polluted water has been developed in recent years.

We have studied the application of the fluidized biofilter bed process (FBBP) as the preliminary treatment for the drinking water supply in the Nan-Tze Waterworks Plant in Shanghai from October 1983 to November 1984.

BASIC PRINCIPLE OF FLUIDIZED BIOFILTER BED PROCESS

The FBBP consists of two main components: the supply of oxygen and the biooxidation process. The flow diagram of the process is to pass raw water through a sprinkling (or trickling) fluidized biofilter bed with supplemental aeration. The supply of oxygen is primarily accomplished by aeration to fulfill the requirements of metabolism for the microorganisms present. At the same time, it can also oxidize other substances that are present in the water. And, in fact, the taste and odor of the water is improved by the FBBP.

Biological oxidation mainly occurs when a great number of microorganisms grow on attached surfaces through their lifetimes. Then the processes of oxidation, reduction, and synthesis by microorganisms can occur, and organic substance are oxidized and changed into simpler organic and inorganic matter. If water contains sufficient amounts of dissolved oxygen and it is enriched with nutrients, biooxidation will occur. Nitrification of ammonia-nitrogen will also take place. When the content of organic substance in the water is found to be comparatively low, then the bionitrification of ammonia-nitrogen will take place first.

EXPERIMENTAL STUDY OF THE FBBP

Objectives

The first problem to be solved for the polluted Huang-pu River is the removal of organic chemicals that consume oxygen. The annual average concentration of ammonia-nitrogen (NH_3–N) is 1 to 2 mg/L. The maximum concentration of NH_3–N is 3 to 4 mg/L, when the water is "pitch-dark and smells." The biochemical oxygen demand (BOD) of this water is 5 to 10 mg/L; chemical oxygen demand (with dichromate) (COD_{cr}) is 20 to 40 mg/L; organic carbon (OC) is 10 to 20 mg/L; total organic carbon (TOC) is 20 to 30 mg/L; and total Kjeldahl nitrogen (TKN) is 2 to 4 mg/L; with a high heavy metals content.

The following experimental studies were completed:

1. The effectiveness of the FBBP for pollution control
2. The relationship between the FBBP and the succeeding treatment process
3. The technical problems of practical application of FBBP for water treatment

EXPERIMENTAL EQUIPMENT

The designed capacity of the model for the intermediate production studies of the FBBP is 1000 m³/day. The equipment is divided into the upper and lower compartments. The upper compartment is a cylindrical body of ϕ 1200 mm diameter × 2000 mm in height, which is a proportion of sprinkling and aeration for the supply of oxygen. It is assembled with a set of 14-pcs nozzles of large diameter and with package material of "vertical wave-type III" 1200 mm in height. The lower cylindrical compartment is of ϕ 1200 mm diameter × 4500 mm height and is filled with the fluidized sand bed medium, which serves as the biocarrier. The particle size of sand is 0.3 to 0.5 mm. The column height is 800 mm. The distribution of water at the bottom is accomplished by a multiple-nozzle system as shown in Figure 1. The relationship between the FBBP and posttreatment processes can be determined by utilizing the dynamic model of water purifying apparatus invented by SMEDI.

THE EFFICIENCY OF THE FBBP

Stage 1 — Development of the biomembrane (or the attached growth of bioslime). The FBBP develops natural bioslimes from the attached growth of microorganisms. This is called "membrane hanging." During the period of "hanging," the hydraulic loading is 35 m³/m²hr. The percentage of bed expansion for the sand layer is 60%.

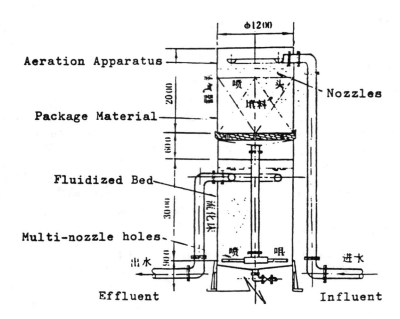

FIGURE 1. The experimental equipment of FBB.

Four indices to evaluate the fluidized bed were selected, i.e., NH_3–N, dissolved oxygen (DO), pH, and turbidity. Raw water temperature during the experimental period was 15 to 18°C. It was found that the raw water contains:

- NH_3–N = 1 to 2 mg/L
- DO = 0.8 to 4.0 mg/L
- pH = 7.0 to 7.5
- Turbidity = 50 to 80 mg/L

After aeration, at the outlet, the DO content of the water is approximately 8 mg/L, which is 80 to 85% of saturation. Dissolved oxygen at the outlet of the FBBP is slightly higher, indicating that oxygen is not completely consumed during this stage. NH_3–N and turbidity do not decrease. After aeration, CO_2 gas escapes, increasing the pH of water at the outlet by 0.4 to 0.5 pH units.

After 3 to 4 weeks, the biofilm is formed and the sand layer has been slightly expanded. The removal of NH_3–N has increased gradually to about 80%. The FBBP starts to consume oxygen, and the value of DO at the outlet remains above 2 mg/L. The raw water turbidity is reduced by 10 to 20%. The pH of the raw water can be raised through the recharge of oxygen, the pH will return to its original value after the FBBP as ammonia-nitrogen is oxidized to nitrite and H^+ ion, and CO_2 gas forms during carbonation.

THE NORMAL OPERATIONAL STAGE

The hydraulic loading of the FBBP is 40 m^3/m^2hr under normal operation. The detention time within the cylindrical body equals 6 min. The sand expansion is 80%. The operating condition of the equipment is steady state. Microscopic examination of the surface of the package material on the vertical wave-type III shows an enriched culture of bacterial slimes, algae (e.g., vorticella), and other plankton. With a cylinder thickness of about 1.5 m of the fluidized sand layer, the biofilms hang on the carriers, and the surface color turns from yellow into brown. The upper layer of the biofilm is very thick, with a heavy growth of vorticella, dendromonas, and a protozoan community. Rotifers and roundworms are also present, but with a lack of bacterial-gelled groups. The lower layer of the biofilm is comparatively thinner, with lack of vorticella and filamentous protozoa, which have less motility. On the other hand, this layer is full of bacteriological slime-organisms from the gelled groups, e.g., "Sphaerotilus".

After the attachment of biofilms, the volume of the particles becomes enlarged, with a corresponding decrease in specific gravity. The exterior shape is basically a sphere. The pressures exerted on its periphery are equal to each other, therefore the hydraulic shearing force or the drag force acting on it will be low. Thus, the drag force does not affect the particle. The quantity and the kind of microorganisms on the carriers vary inversely with the depth of sand layer, i.e., the rule of "declining rate" from above to bottom. This is due to the action of hydraulic separation, so that those floatings on the upper layer are the grown-ripe and those settled underneath are not ripe.

The results of raw water analysis through the preliminary treatment of FBBP are

1. NH_3-N, NO_2-N, and NO_3-N — Ammonia-nitrogen in water is produced from the degradation of protein by bacteria. Factors affecting the removal of NH_3-N are the detention time, temperature, pH, alkalinity, dissolved oxygen, etc. The experimental studies show that the effect of NH_3-N removal mainly depends upon the concentration of ammonia-nitrogen, the dissolved oxygen, and the water temperature, under defined hydraulic loading. The results of ammonia-nitrogen removal are shown in Table 1.

    Table 2 shows the effect of temperature on measurements of NH_3-N, NO_2-N, and NO_3-N. The removal efficiency of NH_3-N is best when its concentration is 0.5 to 2.0 mg/L. It will decrease when the concentration is below 0.5 mg/L due to lack of bacterial nutrients and the fact that a large portion of the residue ammonia compounds are not easily oxidized. When the content of ammonia-nitrogen is greater than 2.5 mg/L, the efficiency of removal will also decrease. This is mainly because of insufficient amounts of dissolved oxygen in the water. The temperature also affects the kind of microorganisms and the speed of metabolism. The best removal is at 20 to 30°C.

Table 1. Percent (%) of NH_3-N Removal

T (°C)	Percent Removal for Different Influent NH_3-N Levels						
	<0.5	0.5—1	1—1.5	1.5—2	2—2.5	2.5—3	3—4
5—15	50	48	43	33	33		
15—20	52	78	77	74	67	70	
20—25	67	90	81	84	75	61	
25—30	64	83	85	80	67	46	50

Table 2. Measurements for NH_3-N, NO_2-N, and NO_3-N at Different Temperatures

	Influent (mg/L)			Effluent (mg/L)		
T (°C)	NH_3-N	NO_2-N	NO_3-N	NH_3-N	NO_2-N	NO_3-N
16	0.60	0.054	0.35	0.10	0.008	0.90
18	0.96	0.115	0.80	0.12	0.012	1.76
20	1.50	0.125	0.96	0.28	0.102	2.17
20	1.68	0.217	0.85	0.34	0.312	2.10
29	2.20	0.075	0.10	1.02	0.200	1.05
30	2.73	0.380	0.50	1.32	0.760	1.50

In the presence of oxygen, nitrification of ammonia-nitrogen produces nitrites and nitrates in two independent stages. When the content of ammonia-nitrogen is high, with low dissolved oxygen, the majority of NH_3-N will remain as nitrites because the bacteria completing nitrification need more oxygen for the conversion to nitrates.

2. BOD_5, COD_{cr}, COD_{KMnO4}, TOC, and TKN — BOD_5, COD, TOC, and TKN are used to understand the oxygen-consumed organic matters in water. Under different temperature conditions, the removal of organic matter by the FBBP is:

BOD_5	27 to 48%
COD_{cr}	24 to 31%
COD_{KMnO4}	10 to 24%
TOC	12 to 22%
TKN	19 to 80%

3. Total Fe, Mn, phenol, Cu, Pb, Zn, and Cd — The raw water of the Huang-pu River is polluted by Fe, Mn, phenols, and heavy metals which directly impair the drinking water quality. Under different temperature conditions, the percentages of removal of these materials by FBBP are as follows:

Total Fe	20 to 28%
Mn	28 to 54%
Phenol	50 to 96%
Cu	30 to 60%
Pb	12 to 50%
Zn	30 to 35%
Cd	20 to 80%

INVESTIGATION OF THE GENERAL CHARACTERISTICS OF THE FBBP

The Influence of the FBBP is Related to Succeeding Unit Processes of Water Treatment

The FBBP has the ability of carrying out biological oxidation, as well as biological flocculation. The average decrease of turbidity of raw water after being treated with FBBP is 16%. Parallel comparative studies have been made with and without the preliminary treatment of the FBBP. The operating parameters were:

- Mixing time = 1.1 min
- Flocculation time = 20.5 min
- Upward velocity of flow through shallow sedimentation basin = 4 mm/sec (length of the inclined pipe = 1 m)
- Rate of filtration = 12 m/hr

Data obtained:

- The average turbidity of settled water via preliminary treatment with FBBP = 16 unit
- The corresponding turbidity of the filtrate = 1.1 unit

While for water without preliminary treatment with FBBP:

- The average turbidity of settled water = 23 unit
- The corresponding turbidity of the filtrate = 1.8 unit

Sedimentation and filtration of the raw water with preliminary treatment by the FBBP improves the water quality when the ammonia-nitrogen content of the raw water is 0.5 to 2.0 mg/L at 15 to 30°C. FBBP decreases NH_3–N and there is a great saving of chlorine for breakpoint chlorination (about 80%) and further savings are obtained as less coagulant is needed.

CHARACTERISTICS OF THE FBBP

Supply of Oxygen

The supply of oxygen and the technique of fluidization are the key points for application of FBBP at the water plant. Ordinarily, in a liquid-solid system, the method of exoteric bed supply of oxygen is used. If air is chosen as the source of oxygen, the degree of saturation of the dissolved oxygen is controlled, and sometimes it will be necessary to recycle the flow in order to fulfill the demand of oxygen supply. It has been determined from experiments that nozzles combined with the package material to form an aeration apparatus can satisfy the requirement of dissolved oxygen for the FBBP, without an increase of the package height, even without any increase of dissolved oxygen. A height of 1200 mm package material will be ample.

Fluidization

The problem of even distribution of flow to the fluidized bed is a focal point of present research. The system has a low head loss, with porous covered plates, and an inverted cone bottom for even distribution of water. Uneven distribution would pile up and cake the carrier, or even more seriously destroy the fluidization. The type of nozzle that is used has an angle not greater than 30°. The range of nozzle type for fluidization varies with the velocity of the nozzle, and it is also found that the area of nozzle to the area of fluidization must fall within a certain ratio. It is not worth the use of a conical bottom for a large-scale fluidized biofilter bed because of the extra increase of the height of the equipment. Through many experimental studies, to minimize the height of the structure, the flat bottom with the accessibility of even distribution and an enlarged zone of influence, which is not easily choked, and a low head loss structure is the best design for the nozzle. Through investigations and measurements, the total head loss of of the fluidized bed will be only 0.8 to 1.1 m when the diameters of the particles are 0.3 to 0.5 mm, and the package height is 800 mm, with a hydraulic loading of 35 to 45 m^3/m^2hr.

Detachment of Biofilms

The shearing forces exerted on the bioparticles are very small after the attachment of biofilms onto the carriers during water treatment. Therefore, the detachment of the biofilms is not a serious problem. During water treatment the growth of biofilms would take time to develop. During the period of study, experience shows the biological carriers are replaced at 10% per month, which would not affect the efficiency of treatment very much. However, if the detachment time is too long, the agglomeration of the particles will spoil the normal operation of the FBBP. Bioparticles drawn from the fluidized bed can be separated from the sand particles mechanically or hydraulically. The carriers can then be recycled to the fluidized bed.

CONCLUSIONS

1. The process of preliminary treatment with a fluidized biofilter bed has a remarkable efficiency for the improvement of raw water quality and has the additional advantages of both bioabsorption and flocculation. The removal of ammonia-nitrogen by the FBPP is very effective. When the water temperature is 15 to 30°C, the ammonia-nitrogen content of raw water is 0.5 to 2.0 mg/L, the average percentage of removal of NH_3–N equals 80%. The FBBP also removes turbidity, organic compounds, and heavy metals.
2. By means of natural sprinkling, filtering, and aeration processes, the FBBP does not need extra equipment for oxygen supply and can basically satisfy the demand of oxygen for nitrification of NH_3–N in raw water sources of the Huang-pu River.
3. The nozzle is designed for the fluidization of the bed material and in a way that improves flow distribution. The head loss of the complete FBBP system is only 3 m.

CHAPTER 21

Removal of Synthetic and Natural Organic Compounds by Biological Filtration

B. E. Rittmann and J. Manem

ABSTRACT

Biological filtration increasingly is being used in drinking water treatment to produce a biologically stable water, to remove synthetic organic chemicals (SOCs), and to denitrify nitrates and nitrites. These filters come in a range of biofilm processes that are operated under highly oligotrophic conditions. Except for dentrification, they are aerobic. The results reviewed here demonstrate that natural organic polymers and several SOCs are substantially biodegraded during passage through a biological filter simulating drinking water treatment. In particular, proteins, phenol, chlorinated phenols, and chlorinated benzenes show significant percent removals immediately or after a short acclimation period.

BIOLOGICAL FILTRATION

Biological filtration is a relatively common practice for preparation of drinking water in Europe,[1-3] and it is beginning to be applied now in the United States and in Japan. Biological filtration is used to achieve three broad goals: (1) biologically oxidize biodegradable components, making the water biologically stable[3] and reducing the need for excess chlorination;[1] (2) biodegrade synthetic organic micropollutants that are harmful to human health;[4] and (3) remove nitrate and nitrite via denitrification.

Many different approaches to biological filtration are being utilized: they include slow sand filters, rapid sand filters, granular activated carbon (GAC) filters, submerged and unsubmerged filters of media larger than sand, rotating biological contactors, fluidized beds, and river bank and dune filtration. Most of these approaches were extensively reviewed recently.[1]

Although the approaches differ substantially in terms of process configuration, they all have two common and critically important characteristics. The first characteristic is that all of the processes are of the biofilm type, in which the bacteria are stably retained in the process by natural attachment to solid surfaces, such as sand, carbon, or rocks. Retention by attachment is a key feature, because the microorganisms are maintained under conditions allowing only very slow specific growth rates. Other traditional retention measures, such as sedimentation, cannot reliably retain cells at these low growth rates. The second characteristic is that the processes select for oligotrophic bacteria or those bacteria especially adapted to function when their substrate concentrations are very low.[5,6] Oligotrophy is important, because organic substrates are present in drinking water supplies at microgram per liter levels. In addition, oligotrophic bacteria are known to consume many different kinds of compounds,[6] the situation prevailing in drinking water supplies.

Except for denitrification processes, most of the biological filtration processes are aerobic, which means that dissolved oxygen is present and utilized as the electron acceptor by the bacteria. This aerobic nature affects which synthetic organic chemicals are subject to biodegradation in a biological filter used in drinking water treatment. In general, SOCs that are built upon an aromatic structure should be most susceptible to biodegradation, while halogenated aliphatics should be most resistant.[4,7]

The purpose of this paper is to provide a state-of-the-art review of the biodegradation capacity of aerobic biological filters for natural organic polymers, which comprise the majority of the organic matter in drinking water supplies, and commonly found SOCs. The bulk of the material comes from the recent work of Manem.[8]

REMOVAL OF NATURAL ORGANIC POLYMERS

Most of the dissolved organic matter in natural waters is composed of organic polymers, such as humic and fulvic acids, polypeptides, and carbohydrates. The conventional wisdom about these natural organic polymers in surface waters and ground waters is that they are resistant to biodegradation. However, recent information is demonstrating that the natural organic polymers are biodegradable and can support substantial microbial growth in distribution systems (a serious problem) and in biological filters (a desired situation).

The first evidence that natural organic polymers are relatively biodegradable came from new methods to measure the low concentrations of biodegradable organic matter (BOM) in waters.[9,10] These methods found that 11 to 59% of the DOC in raw waters in France was biodegraded. Absolute concentrations of biodegradable organic carbon were greater than 200 µg C per liter before biological filtration.

A laboratory study[11] on biological filtration of natural humic materials showed that about 10% of the humic material (as C) was removed, but that substantial biofilm accumulation occurred. The biofilm was capable of removing two natural odor-causing chemicals — geosmin and methylisoborneol — as well as the industrial odor-causing chemicals phenol and naphthalene.

The studies on BOM measurement and biological filtration of humic materials demonstrate two factors about natural organic polymers: they are biodegradable and can support biofilm accumulation. One implication is that these polymers can cause significant water-quality problems if allowed into the distribution system: namely tastes, odors, turbidity, corrosion, and plate counts.[1] The second implication is that the natural polymers can support good biofilm accumulation in a biological filter. Good biofilm accumulation means that the humic materials and other compounds, including taste-and-odor components and SOCs, can be removed during filtration.

Manem[8] investigated in detail the fate of two classes of high molecular weight organic polymers in a laboratory-scale biological filter designed to simulate a full-scale filter at Croissy, France. Figure 1 is a schematic drawing of the laboratory filter, and Table 1 presents design parameters for the filter. The design, which was based on biofilm kinetics, successfully simulated the nitrification reaction in the full-scale filter.[8,12] The two classes of compounds were polysaccharides and proteins.

The polysaccharide was dextran, a branched polymer of D-glucose with an average molecular weight of 40,000. The proteins were lysozyme and β-lactoglobulin. Lysozyme is an enzyme with a molecular weight of 14,600 and a net positive charge. β-Lactoglobulin has a molecular weight of 18,400 and a net negative charge at neutral pH. ^{14}C-Methylated forms of dextran and both proteins were available and made it possible to assay concentrations of the original compound and CO_2 by scintillation counting.[12]

Figure 2 presents the fates of dextran, lysozyme, and β-lactoglobulin, as well as CO_2 productions, during short-term (<1 day) exposures of these chemicals in the influent of the biological filter. Table 2 summarizes the steady-state removals and CO_2 productions.

Dextran shows an apparent removal of 11%, with 6% of the C oxidized to CO_2. However, these effects do not reflect biodegradation of dextran; instead, low molecular weight impurities were degraded and accounted for the transformations.[8] The lack of degradation of dextran in short-term experiments probably was caused by a lack of all the hydrolytic enzymes needed to break the bonds

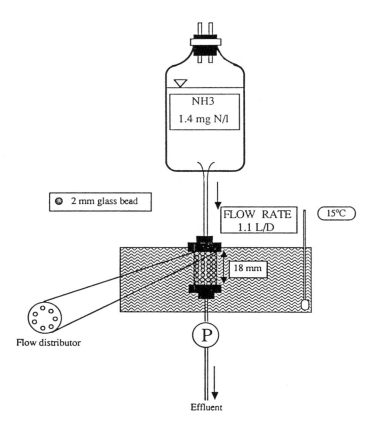

FIGURE 1. Schematic of the laboratory-scale biological filter used to simulate the full-scale filter at Croissy.

Table 1. Design Parameters of the Laboratory-Scale Biological Filters

Parameter	Value
Medium	Glass beads
Medium diameter	2 mm
Medium depth	18 mm
Superficial flow velocity	6.2 m/day
Porosity	0.38
Influent concentration	
NH_4^+–N	1.4 mg/L
Acetate	0.2 mg/L

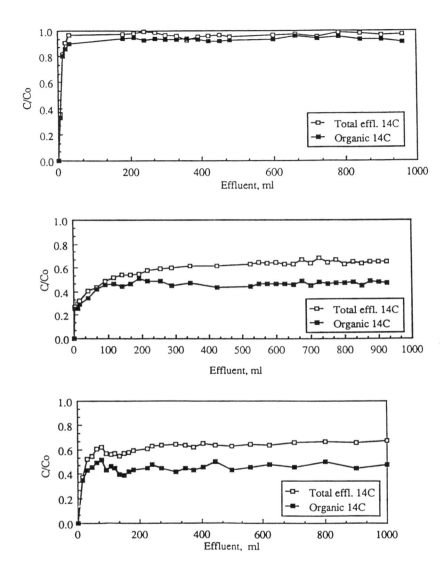

FIGURE 2. Fates of (a) dextran; (b) lysozyme; and (c) β-lactoglobulin in short-term applications to the biological filter. Input concentrations were 200 μg/L.

Table 2. Summary of Percent Removals of Input Compound and CO_2 Productions According to ^{14}C Analyses

	% Removal	% CO_2 Formed
Dextran	11	6
Lysozyme	53	18
β-Lactoglobulin	54	18

between the glucose monomers. Because polysaccharides are important in the structures of cell walls, membranes, and capsules, the degradation of these polymers must not be accomplished easily by unacclimated bacteria. The biological filters, which had been exposed primarily to NH_4^+ and acetate, were not able to hydrolyze the dextran bonds without much more extensive acclimation.

Lysozyme, a positively charged protein, was hydrolyzed (53% removal) and mineralized (18% CO_2 production). There was a net buildup of smaller molecules (mol wt $\leq$500), which indicates hydrolysis of the large protein to smaller subunits.[8] However, the CO_2 formation demonstrates at least partial mineralization of the hydrolysis fragments.

β-Lactoglobulin, a negatively charged protein, exhibited removal (54%) and mineralization (18%) very similar to those of lysozyme. Also, small molecular weight fragments accumulated, indicating that the rate of hydrolysis exceeded the rate of mineralization.[8]

The much greater extent of hydrolysis and mineralization for the proteins, compared to the polysaccharide, occurred because amino acid monomers of proteins are linked by a common peptide bond. Thus, the protease enzymes available in almost all bacteria are able to hydrolyze any protein. The proteolytic enzymes present at the surface of the biofilms grown primarily on acetate and NH_4^+ were able to hydrolyze two proteins about equally without an acclimation period.

In summary, a substantial portion of the natural organic polymers in waters is biodegradable. Included are humic materials and proteins. Polymers having many different linkages, such as dextran, appear to be more resistant to hydrolysis, at least without extensive acclimation.

REMOVAL OF SYNTHETIC ORGANIC COMPOUNDS

Aerobic bacteria are able to transform many of the common SOCs. Although the 1- and 2-carbon chlorinated hydrocarbons usually are refractory to aerobic bacteria, benzenes, phenols, many pesticides, and low molecular weight petroleum constituents usually can be biodegraded aerobically.[1,4] Hence, significant potential exists for biological removal under the conditions of biological filtration in drinking water treatment.

Although biodegradation potential exists, its realization is still largely unknown and poorly predictable. One reason for the poor prediction is that basic kinetic data are missing for most of the SOCs under conditions relevant to biological filtration. Those relevant conditions include microgram-per-liter concentrations of the SOCs, generally oligotrophic concentrations for all substrates, and biofilm retention.

One concept that must be kept in mind when interpreting transformation kinetics of SOCs is that they probably behave as secondary substrates. A secondary substrate is a compound whose biodegradation provides zero or negligible energy for cell growth and maintenance.[13] Secondary substrates typically are present at very low concentrations or transiently. The biomass that degrades the secondary substrate is supported by the degradation of a more plentiful primary substrate, which can be a single compound or the sum of many secondary substrates. In biological filtration, the SOCs probably are secondary substrates, and the natural organic polymers comprise the bulk of the primary substrate.

Manem[8] carried out research to assess the removal kinetics for phenol, five chlorinated phenols, and three chlorinated benzenes. The same laboratory-scale biofilm reactors were used to simulate field conditions at the Croissy plant. Input concentrations from 1 to 400 µg/L were applied for short-term (1 hr to 1 day) experiments. The concentrations of chlorophenols and chlorobenzenes were measured by gas chromatography after hexane extraction.[8] ^{14}C-Phenol and -chlorobenzene also could be assayed by scintillation counting.[8]

Figure 3 presents the ^{14}C results for phenol and chlorobenzene. Phenol was more than 90% removed from the beginning of the experiment, and about 30% of the ^{14}C was recovered as CO_2. The low recovery of ^{14}C (about 40%) is expected and can be explained by the incorporation of C from phenol into newly formed biomass. The removal pattern for chlorobenzene was quite different. Initially, little or no removal occurred, and a chlorobenzene breakthrough resulted for the first 3 hr. Then, an acclimation took place from 3 to 8 hr, after which biodegradation increased rapidly, $^{14}CO_2$ appeared in the effluent, and a steady-state removal was approached. Thus, chlorobenzene was substantially biodegraded in the laboratory-scale filter, but a short acclimation period (<1 day) was needed. The length of the acclimation period suggests that enzyme derepression had to be triggered by the presence of chlorobenzene, but the biofilm microorganisms possessed the capability for degrading chlorobenzene.[8]

The patterns for o-chlorophenol and 2,4- and 2,6-dichlorophenol were similar to the pattern of phenol: no acclimation period was observed. On the other hand, 1,2- and 1,4-dichlorobenzene showed acclimation periods of about 8 hr before rapid removals commenced.

Table 3 summarizes the removal of all the compounds during the steady-state phase. For the lowest input concentrations, significant removals occurred for phenol, chlorobenzene, the dichlorobenzenes, chlorophenol, and the dichlorophenols. Removal of the trichlorophenols could not be distinguished from zero. These results are significant, because they demonstrate that the biodegradation kinetics are fast enough under conditions of biological filtration that several of the most common aromatic SOCs are removed to a major degree.

Comparison of the removals for different input concentrations reveals that the highest percentage removals always occurred for the lowest input concentrations. This trend is expected, because higher concentrations should begin to saturate the microbial reaction. Thus, the best percentage removals are expected for the micropollutant concentrations frequently detected.

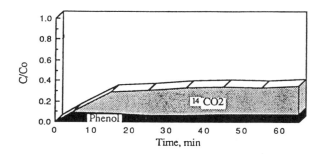
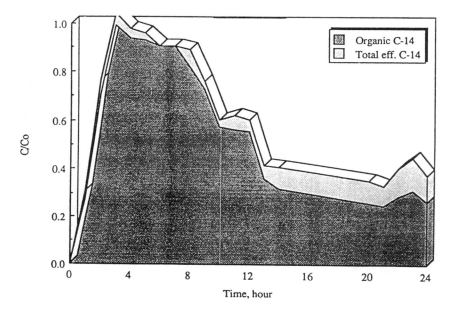

FIGURE 3. Effluent/influent concentration profiles for original compound and $^{14}CO_2$ when ^{14}C-labeled compounds were used: (top) phenol at 2.2 μg/L; (bottom) chlorobenzene at 2 μg/L.

SUMMARY

Biological filtration increasingly is being used in drinking water treatment to produce a biologically stable water, to remove SOCs, and to remove nitrates and nitrites. These filters are biofilm processes operated under highly oligotrophic conditions. Except for denitrification, they are aerobic.

The results reviewed here demonstrated that natural organic polymers and synthetic organic chemicals are substantially biodegraded during passage through

Table 3. Summary of Steady-State Removal Percentages for SOCs in the Laboratory-Scale Biological Filter

Compound	Input Concentration (μg/L)	Percent Removal
Phenol	2.2	92
	20	86
	400	17
o-Chlorophenol	10	85
2,4-Dichlorophenol	1	30
2,6-Dichlorophenol	1	24
2,4,6-Trichlorophenol	1	<2
2,3,4-Trichlorophenol	1	<2
Chlorobenzene	2	75
	10	75
1,2-Dichlorobenzene	1	81
	5	69
1,4-Dichlorobenzene	1	73
	5	68

a laboratory-scale filter that simulates a full-scale biological filter in France. In particular, proteins, phenol, chlorinated phenols, and chlorinated benzenes showed significant percentage removals as secondary substrates fed to filters whose biofilms had been grown through the oxidation of NH_4^+ and acetate. A short acclimation period was required for the chlorinated benzenes. Dextran, a polysaccharide, was not degraded during a short-term test, probably because the biofilm bacteria did not have the range of specialized hydrolytic enzymes needed to cleave the polymer into glucose monomers.

The key finding is that several important components in water supplies are significantly biodegraded under conditions appropriate to biological filtration in drinking water treatment.

REFERENCES

1. Rittmann, B. E., and P. M. Huck. "Biological Treatment of Public Water Supplies," *Crit. Rev. Environ. Control* 19:119-184 (1989).
2. Bouwer, E. J., and P. B. Crowe. "Biological Processes in Drinking Water Treatment," *J. Am. Water Works Assoc.* 80(9):82-91 (1988).
3. Rittmann, B. E., and V. L. Snoeyink. "Achieving Biologically Stable Drinking Water," *J. Am. Water Works Assoc.* 76(10):106-114 (1984).
4. Rittmann, B. E. "Biological Processes and Organic Micropollutants in Treatment Processes," *Sci. Total Environ.* 47:99-113 (1985).
5. Rittmann, B. E., L. Crawford, C. K. Tuck, and E. Namkung. "In situ Determination of Kinetic Parameters for Biofilms: Isolation and Characterization of Oligotrophic Biofilms," *Biotechnol. Bioeng.* 28:1753-1760 (1986).
6. Roszak, D. B., and Colwell, R. R. "Survival Strategies of Bacteria in the Natural Environment," *Microbio. Rev.* 51:365-379 (1987).

7. Rittmann, B. E., D. Jackson, and S. L. Storck. "Potential for Treatment of Hazardous Organic Chemicals with Biological Processes," in *Biotreatment Systems,* Vol. III, D. L. Wise, Ed. (Boca Raton, FL: CRC Press, Inc., 1988), pp. 15-64.
8. Manem, J. "Interactions Between Heterotrophic and Autotrophic Bacteria in Fixed-Film Biological Processes Used in Drinking Water Treatment," PhD Dissertation, University of Illinois, Urbana, IL.
9. Servais, P., G. Billen, and M. C. Hascoet. "Determination of the Biodegradable Fraction of Dissolved Organic Matter in Waters," *Water Res.* 21:445-450 (1987).
10. Joret, J.-C. "Rapid Method for Estimating the Bioeliminable Carbon in Drinking Water Treatment Plants," presented at the Annual Conference of the American Water Works Association, Orlando, FL, June 1988.
11. Namkung, E., and B. E. Rittmann. "Removal of Taste and Odor Compounds by Humic-Substances-Grown Biofilms," *J. Am. Water Works Assoc.* 79(7):107-112 (1987).
12. Manem, J., and B. E. Rittmann. "Scaling Procedure for Biofilm Processes," Proceedings of Technical Advances in Biofilm Reactors, Nice, France (April 1989), pp. 343-360.
13. Namkung, E., R. G. Stratton, and B. E. Rittmann. "Predicting Removal of Trace-Organic Compounds by Biofilms," *J. Water Pollut. Control Fed.* 55:1366-1372 (1983).

CHAPTER 22

Membrane Filtration in Drinking Water Treatment: A Case Story

J. Mallevialle, J. L. Bersillon, C. Anselme, and P. Aptel

SUMMARY

This paper presents all the aspects of pilot-plant experiments and industrial implementation of a new technique to clarify water: ultrafiltration. This technique is assessed with respect to the quality of the resulting water in a case of a resource displaying a very variable quality. It is shown that the treated water has a very good constant quality, especially as far as turbidity and correlated quality parameters are concerned.

INTRODUCTION

Small systems have problems of their own. Most frequently, they provide water in regions where the population is very scattered. Therefore they cannot be avoided, as the connection of houses to a major network would represent unaffordable expenses in piping. Small communities rely then on local water resources, and most often, the treatment applied to these resources is less than adequate for at least one of the following reasons:

1. In the case of community-owned systems, the municipality cannot afford to have a water specialist to take care of the treatment and the resulting water quality is not always irreprochable.
2. The water resource may display quality changes incompatible with the water treatment system.

In this case, when the resource quality deteriorates, the distributed water quality deteriorates as well to the point where it does not comply with quality standards anymore.

The site of Amoncourt used to be such a case. Amoncourt is a small community of 320 inhabitants, located in Eastern France, within the karstic belt of the Parisian Basin. Its water resource is constituted by a spring. It is marred by occasional turbidity and total organic carbon (TOC) spikes which can reach up to 300 NTU and 9 mg/L,[1] respectively, within a few hours, whereas its normal turbidity ranges 2 to 5 NTU.

The old plant was based on direct filtration, which resulted in strong turbidity spikes in the distributed water, as such a process is not normally meant to handle such raw water qualities. It was actually found that this water resource should have been equipped with a complete clarification system (coagulation-settling-filtration) which would have been quite difficult to manage in all cases. In such a situation, it was estimated that the population was spending 700 FrF per household per year in bottled (mineral) water to provide for the potable water needs. The peak water demand was 180 m^3/day.

THE ALTERNATIVE TREATMENT: ULTRAFILTRATION

Among pressure-driven membrane processes, ultrafiltration is situated between microfiltration, which stops particles down to 0.1 μm, and reverse osmosis, which stops ion-sized particles. It is a process mostly used in food processing and pharmacy to filter or concentrate liquids.

In water treatment, one of the most important problems to solve is the separation of colloids from water, which most of the time involves the addition of chemicals (coagulants, filter acids, flocculants). These reagents must be added to the water at a rate depending on the water quality itself. In membrane processes, this addition is not necessary, provided that the membrane has a porosity tight enough to stop the unwanted particles. Figure 1 summarizes the water pollutants and the range of action of the pressure-driven membrane processes.

In other words, provided that the membrane porosity (cut-off threshold) is properly chosen, it will do the work normally assigned to clarification, without requiring reagents for the separation. These considerations lead Lyonnaise des Eaux/Dumez (France) to choose the ultrafiltration process to treat resources like Amoncourt.[1]

FROM FEASIBILITY TO PLANT SPECIFICATIONS

The feasibility study aimed at the answers to the following questions:

- Which type of membranes should be used?
- Which process should be implemented?

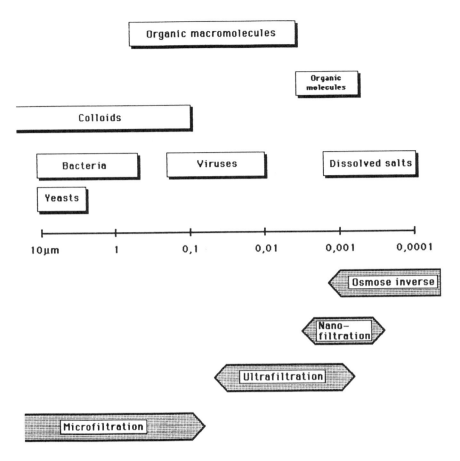

FIGURE 1. Summary of the membrane cutoffs (lower part) and the size of pollutants (upper part).

These two questions were addressed through two pilot-scale campaigns:

- The first one to compare membranes of different natures on a same raw water quality and to determine a range of feasable process conditions.
- The second campaign to fine tune the operating conditions and to assess the overall behavior of the system (membrane and process) under stress conditions (turbidity spike).

Table 1. Operating Conditions for the Choice of the Membrane

Membrane ID	Cutoff (μm)	Fiber diameter (mm)	Tangential velocity (m/sec)	Operating (bars)
Ceraver-SCT	0.2	7	0.71	1.2
			1.40	1.2
Lyonnaise	0.01	1	0.5	0.5
des Eaux/Dumez			0.75	0.5
			2.5	1.2

Choice of the Membrane Nature

For the Amoncourt case, two possible membranes were chosen from previous experiments:

- A ceramic membrane (Ceraver-Tarbes, France)
- An organic membrane (BRO23, Lyonnaise des Eaux/Dumez, Toulouse, France)

These membranes were selected among a wider range of membranes including commercial products (Enka AG, Romicon, etc.). The results of these experiments will be published separately.

These membranes were placed on a small-scale pilot unit, allowing for pressure and flow measurements, and backwashing. The range of operating conditions is summarized in Table 1.

From these experiments, the following results were drawn:

1. The organic membrane (BRO23) is preferable to the ceramic membrane because of its better ability to be backwashed, hence to make it possible to maintain a feasable flux for a long period of time.
2. For this membrane, the feasable water flux was estimated at 100 L/hr/m at an operating pressure of 0.5 bars. These data were used as a starting point for the validation experiments carried out on a larger scale module.
3. For the production, the specific energy required by the filtration amounts to 85 Whm^{-3} (306 KJm^{-3}). This was calculated from the headloss due to the tangential velocity (recirculation) and to the pressure loss across the membrane. For the ceramic membranes, the specific energy ranged 400 to 550 Whm^{-3}.
4. The filtered water from the organic membrane has a turbidity lower than 0.1 NTU. This offers the possibility to comply with the most stringent standards (U.S. EPA Recommendation for Safe Drinking Water Act).
5. Plate counts performed on the filtered water showed no evidence of microbial breakthrough.

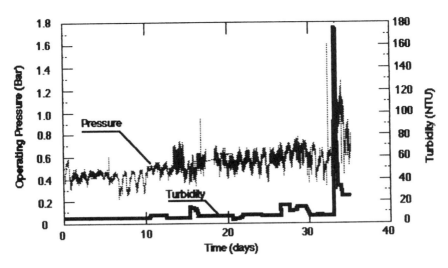

FIGURE 2. Raw water turbidity and operating pressure during the validation experiment.

Validation and Operating Conditions

For the validation experiments, a medium scale (MS) module was mounted on a computer-controlled pilot unit. The MS module has 5 m² of filtering surface area constituted by BRO23 membranes. The pilot unit is equipped so as to control the process (pressure and flow rate, temperature) and to record process data for later treatment.

The hydraulic conditions used in this experiment were set so as to produce a constant permeate flux. It was chosen to have a constant throughput in order to simplify the dimensioning of the production unit. Under this condition, the membrane fouling gives rise to an increase in the operating pressure. The periodic backwashing makes it possible to restore the operating pressure to its nominal value. This is illustrated by Figure 2, where it can be seen that the operating pressure is nearly constant with time, when the raw water turbidity is less than 25 NTU. When the turbidity is higher, the operating pressure tends to rise, as illustrated by Figure 3.

In a 1-month period, it was confirmed that the filtered water turbidity remained lower than 0.1 NTU, regardless of the raw water turbidity. One must keep in mind that this was accomplished without the help of any reagent for the separation.

Organic parameters such as TOC, UV absorbance at 254 nm, and trihalomethane formation potential (THMFP) were also measured during this experiment. It was found that the TOC was removed by about 20%, whereas the UV absorbance and the THMFP were removed by about 40%. This is attributed to

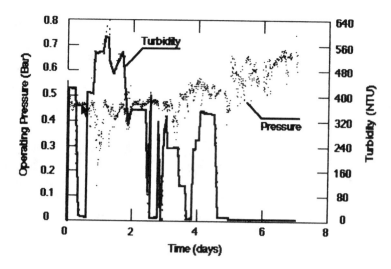

FIGURE 3. Operating pressure response to high turbidity spike (validation experiment).

the contribution of particulates and colloids to these later qualitative parameters (UV and THMFP), whereas the TOC contribution related to these materials is lower. Also, it was noticed that the turbidity spikes are accompanied by TOC spikes, which are better removed than the "background" TOC. This is illustrated by Figure 4, where the background TOC is estimated to be 2 mg/L, whereas TOC spikes can reach 8 to 10 mg/L.

The Plant Specifications

The required peak throughput was set to 10 m^3/hr. In order to comply with this requirement, the actual plant design involved two independant lines of 10 MS modules in parallel. The operating conditions were unchanged with respect to those used in the validation pilot experiments, as they gave satisfaction. The plant was equipped with a programmable automaton, in charge of managing the different operations such as flow rate controls, pressure monitoring and control, and backwash sequences.

PLANT STARTUP AND OPERATIONS

Membrane Manufacturing and Water Treatment

The membrane manufacturing, as well as all the steps leading from a membrane to the storage requirements for the modules, involves the use of specific materials

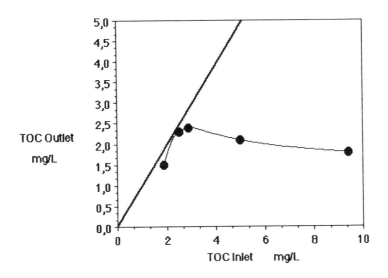

FIGURE 4. Outlet TOC as a function of inlet TOC during a turbidity spike. The solid line represents 0% removal.

and reagents, insuring tightness, material compatibility (glueing), and preservation of the membrane and module properties from deterioration through aging. In addition to these requirements on materials and reagents, the use of this equipment in potable water treatment involves other requirements on the nature of the materials, as well as the nature and concentration of specific solubles admissible for drinking water, disposal into a sewer, or disposal into the environment.

This situation may end up in a conflict of requirements, which in the present case was solved because there has been a constant link between the manufacturer and the end user. This resulted in the choice of materials and reagents which in turn involved a fine tuning of the membrane and module manufacturing processes. Also, the startup of the plant has been studied in detail at the laboratory scale, by the assessment of the leaching out of substances contained in the conditioning solution of the modules.

The outlet concentrations of the undesirable substances were modeled by a relationship of the kind:

$$C(t) = aC_o/[1 + kV.s(t)]^2 \qquad (1)$$

where $C(t)$ = time-dependent concentration

C_o = initial concentration

$V(t)$ = time-dependent volume of leachate

a, k = adjustable constants

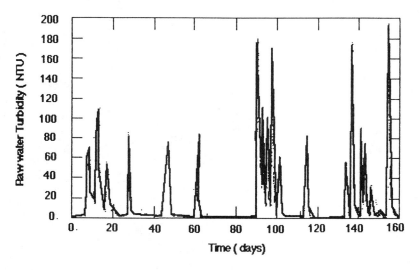

FIGURE 5. Turbidity profile of the Amoncourt resource for the first 6 months of operation.

The relevant pollutants assessed by this method were

- TOC, BOD5, COD
- Specific chemicals used as:
  Membrane additives
  Preserving agents
  Wetting agents
  Bacteriostatic agents

The comparison of the leaching law (1) with the different standards for disposal and drinking water production led to the recommendations of modifications in the module conditioning solutions, in order to minimize the required volume of leachate necessary to comply with these standards.

Again, this was made possible by the very strong links between the manufacturing team and the drinking water technicians. The recommendations were applied to the actual plant startup and were entirely satisfactory.

Plant Operation

Started in November 1988, the plant was kept under close scrutiny for the first 6 months of operation. The reaction of the plant to the variations of raw water quality was examined with respect to the process parameters as well as the treated water quality. Figure 5 illustrates the turbidity profile of the resource

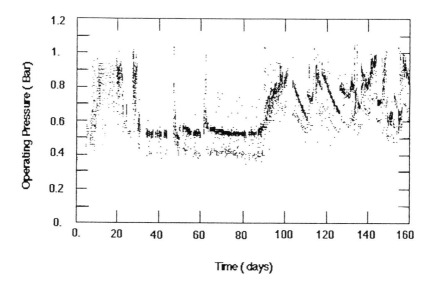

FIGURE 6. Operating pressure profile of the Amoncourt plant and trihalomethanes in the ultrafiltered chlorinated water (THM treated W).

during this period. It can be seen that turbidity spikes (100 to 180 NTU) were encountered, corresponding to rain episodes, whereas the "background" turbidity ranged 2 to 5 NTU. The response of the plant to these variations is illustrated by Figure 6, where the operating pressure is recorded as a function of time as the water throughput is kept constant. As in the case of the pilot plant, this operating parameter tends to rise with rising turbidity.

It is interesting to notice that the overall process makes it possible to "rinse out" the system once the turbidity spikes are over. This happens in most cases, although it requires more time for the pressure than for the turbidity to drop to their nominal values. The plant required only two module regenerations during this period (at 30 and 130 days of operation). During this whole period, the treated water turbidity ranged 0.1 NTU, therefore consistent with the pilot-plant results.

Since ultrafiltration is a new process involving new kinds of equipment, the produced water must be checked with respect to drinking water quality standards by means of grab sampling and analysis by an officially approved laboratory (Ministery of Health). Among all the quality parameters which are measured for this purpose, the following selection is worth considering.

1. The microbial counts, which are illustrated by Table 2 — As suspected during the pilot-plant experiments, there is no microbial breakthrough in the ultrafiltered water: these measurements were duplicated and performed by both the LdE analytical laboratory and the regional laboratory of the Ministry of Health.

Table 2. Microbial Quality of the Resource (RW) and the Ultrafiltered Chlorinated Water (TW) on the Site of Amoncourt

	Sampling date							
	01/03		02/14		03/14		04/11	
	RW	TW	RW	TW	RW	TW	RW	TW
Coliforms (100 mL)	25	0	4	0	3	0	140	0
Fecal coliforms (100 mL)	0	0	1	0	0	0	92	0
Spectro (100 mL)	25	0	1	0	1	0	38	0
Clostridium (100 mL)	5	0	ND	ND	8	0	60	0
Total germs (1 mL)								
37°C	16	<5	18	<5	34	<5	108	<5
22°C	322	<5	81	<5	233	<5	360	<5

Table 3. Al and Fe Contents

Sampling Date	Raw Water			Treated Water		
	Al (μg/L)	Fe (μg/L)	pH	Al (μg/L)	Fe(μg/L)	pH
01/03	216	20	7.46	2	BDL	7.59
02/14	136	10	7.39	15	BDL	7.41
03/14	90	30	7.14	6	BDL	7.14
04/11	430	140	7.46	2	BDL	7.25

Note: BDL: below detection limit.

2. Aluminum and iron — These two elements are present in the resource, and it was found that the iron content was correlated to the turbidity. Table 3 summarizes the raw water and treated water contents for these two elements:

    The removal of these elements by ultrafiltration is attributed to their occurrence as colloidal hydroxides, as these two elements are known to precipitate in such a way. A number of authors presented evidence of a dependency between measured aqueous aluminum and iron and the filtration cutoff. Although surprising at first sight, the removal of these elements and perhaps manganese can be explained by the very low cutoff (0.01 μm) of the membranes. Based on this explanation, it is believed that microfiltration (0.2 μm and above) would allow for a breakthrough of these elements, although there is no documented evidence of this phenomenon.

3. THMFP in the raw water and THM in the treated water — Since the treated water is chlorinated prior to its storage and distribution, and since the raw water is not prechlorinated prior to its treatment, the comparison of these two pieces of information gives a good estimate of the removal of the substances involved in the formation of THMs. This is illustrated by Figure 7. It can be seen that the removal of THMFP is almost complete. This is not in accordance with other results which suggest a removal in the area of 40%. This is attributed to milder chlorination practice in drinking water, as compared to the evaluation of THMFP as performed in laboratory conditions.

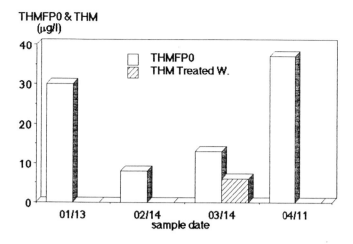

FIGURE 7. Trihalomethane formation potential of the resource (THMFP) and trihalomethanes in the ultrafiltered chlorinated water (THM treated W).

CONCLUSIONS

This case is to the authors' view a typical case of a new process put to work in an industrial situation. The very thorough approach does not make any room for short cuts, leading to an industrial process which still needs to be fine tuned: the plant works currently in nominal conditions.

Technically it was shown that ultrafiltration makes it possible to solve the problems brought about by resources marred by very variable quality, especially with respect to turbidity and correlated quality parameters (Al, Fe, microbials). Also, it is of utmost importance that the onsite experiments have a long enough duration so as to assess properly the specifications of the plant. The risk consists in missing "crisis" situations and cumulative effects which are not yet predictible with respect to their consequences on the operating conditions and the resulting treated water quality.

This alternative treatment is believed to be the most readily available technique to make it possible to comply with the most stringent drinking water quality standards.

ACKNOWLEDGMENTS

The Research and Development of this work have been partially funded by the European Community Commission (BRITE Project RI1B-0155) and the French Ministry of Research (EUREKA Project EU5).

REFERENCE

1. Bersillon, J. L., "Fouling analysis and control," in *Future Industrial Prospects of Membrane Process*. Cecile, L. and Toussaint, J. C., Eds., Elsevier Applied Science, London (1989).

CHAPTER 23

Ultrafiltration of Lake Water: Optimization of TOC Removal and Flux

M. M. Clark

OBJECTIVES

The objectives of this study are to examine batch and semicontinuous ultrafiltration for the treatment of water from Lake Decatur, IL, a public water supply in the midwestern United States. Quality of treatment is judged based on removal of turbidity, total organic carbon (TOC), trihalomethane formation potential (THMFP), and ultraviolet (UV)-absorbing materials from the water. Treatment efficiency is judged on the ability to prevent the loss of membrane flux. Mechanisms of membrane fouling were studied using gel permeation chromatography and static adsorption tests.

INTRODUCTION

As opposed to recent U.S. EPA-funded work in Florida on the treatment of very high TOC, brackish surface and ground waters with reverse osmosis membranes,[1] the present work examines a potential alternative technology for the treatment of typical surface water from the midwestern United States. We are therefore proposing a technology which can be considered in competition with fairly successful classic surface water treatment technologies (i.e., coagulation/flocculation, softening). To be competitive with these classic technologies, relative advantages of the proposed new technology must be demonstrated. These advantages could well emerge in the United States due to the higher turbidity, microbiological, THM, and other disinfection by-product (DBP) standards being contemplated by the U.S. EPA. An ultrafiltration plus coagulation-based treatment system has the further advantages of a smaller flocculation tank (since large floc are not required) and the total elimination of settling and deep-bed filtration stages. Turbidity spikes associated with conventional filter breakthrough, channelizing, and startup after backwash could be totally eliminated with a membrane-based treatment system. Compared to reverse osmosis, ultrafiltration has the advantage of higher flux (because of the larger UF pore size), lower pumping costs (lower pressure required in UF), smaller membrane area, and no brine disposal costs — the residual from ultrafiltration of coagulated lake water is actually a dense cake layer. The frequency of backflushing for ultrafiltration of coagulated waters would be more similar to backwash frequency of conventional deep-bed filters (twice daily). UF membrane *cleaning* could be a less frequent process employing a detergent, base, or other cleaner.

EXPERIMENTAL PLAN/APPROACH

Water from Lake Decatur was transported to the University of Illinois laboratory, prefiltered through a large pore paper filter, dosed with a bactericide, and stored at 4°C. These precautions were developed in order to ensure a stable TOC during sample storage in the laboratory. Since the raw water turbidity was decreased about 25% by the prefiltration process, this water was "reconstituted" with 1 mg/L kaolin clay prior to ultrafiltration experiments in order to achieve the same turbidity as the raw lake water.

Two membrane filtration systems were used in this work. The first system is the so-called "dead-end" batch system which can filter a 250-mL sample (Figure 1). A hollow fiber ultrafiltration system was also used (Figure 2). This system is based around Amicon hollow-fiber membrane cartridges. Coagulation pretreatment was accomplished in the 20-L Rushton vessel. This configuration was used in batch and semicontinuous modes. Semicontinuous operation was achieved by returning retentate and permeate to the 20-L vessel and periodically replacing the entire contents with fresh solution.

TOC, UV, turbidity, and THMFP measurements were made on permeate samples. Particle size measurements were made on the raw and reconstituted waters.* Gel permeation chromatography measurements on raw, coagulated, and powdered activated carbon (PAC)-treated samples were also made in order to determine the molecular weight of the likely organic foulants.

---

* Particles could not be detected in permeate samples using either the electrical sensing zone or photon correlation spectroscopy techniques.

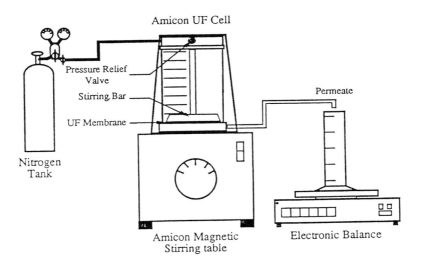

FIGURE 1. Amicon dead-end batch filtration assembly.

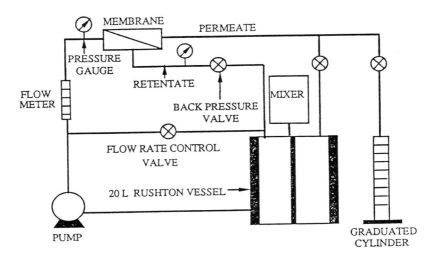

FIGURE 2. Hollow fiber ultrafiltration system.

Static adsorption tests were conducted on fresh membranes to estimate the importance of a pure adsorptive fouling mechanism. In these tests, clean membranes were brought in contact with lake water, under conditions in which there was no flow through the membranes. After exposing the membranes to the lake water in this manner for a fixed amount of time, clean water permeability through the membrane was determined.

Table 1. Survey of Characteristics of Membranes Used in this Work[8-10]

Membrane Class	Description	Available Molecular Weight Cutoff (MWCO)
PM	Hydrophobic polysulfone membrane, negatively charged below pH 8. High flow membrane, which does not adsorb ionic or inorganic solutes, but may adsorb steroids and hydrophobic macromolecules.	10–30 K
XM	Hydrophobic acrylic copolymer membrane, negatively charged. Recommended for concentration of large macromolecules or concentration of particles from molecular material.	50–300 K
YM	Hydrophilic regenerated cellulose membrane, which is probably positively charged. Demonstrates low specific binding with proteins. Has sharp molecular weight cutoff characteristics.	2–100 K

Modified from *J. AWWA*, Vol. 81, No. 11 (November 1989). By permission of American Water Works Association.

Table 2. Organic and Particle Fractionation by Membranes

	TOC (mg/L)	UV	Turbidity (NTU)
Reconstituted lake water	6.14	1.621	4.90
Nuclepore (0.45-μm filter)	5.04	1.086	0.48
Millipore (0.22-μm filter)	4.52	0.922	0.12
XM 100	3.55	0.710	0.05
YM 100	3.53	0.821	0.05
PM 30	3.24	0.730	0.05
YM 5	3.60	0.725	0.05

RESULTS

Table 1 show the results of a literature study to determine the physical/chemical characteristics of the Amicon membranes used in our work.

We examined the organic composition of the lake water using batch membrane separation techniques and gel permeation chromatography. Table 2 shows organic and particulate fractionation using different membranes and the Lake Decatur water. Interestingly for this water, there is little effect on TOC removal as we decrease the membrane molecular weight cutoff (MWCO) below 100 K: TOC and UV removals are about the same for the 100-K, 30-K, and 5-K MWCO membranes. Our gel permeation chromatography (GPC) results explain this phenomenon very nicely (Figure 3): a large fraction of the organic carbon in the lake water has an average molecular weight around 1 K, while a somewhat smaller fraction has a molecular weight larger than 100 K (the exclusion peak). Therefore, a 5-K MWCO membrane really shouldn't be expected to be more efficient than a 100-K MWCO membrane in removing TOC since both

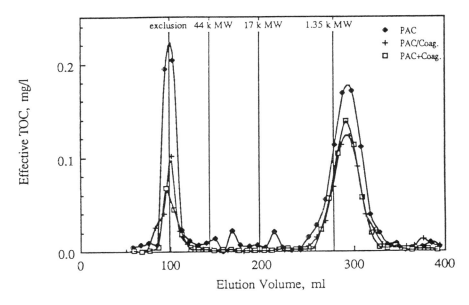

FIGURE 3. Molecular weight distribution of PAC-adsorbed, and PAC-adsorbed and coagulated Lake Decatur water.

membranes would pass the lower molecular weight fraction, while both membranes would retain the higher molecular weight fraction. Evidently, pretreatment before UF is necessary to achieve really significant TOC removals. Figure 3 shows the relative greater effectiveness of PAC over coagulation for the removal of the low molecular weight compounds and the relatively greater effectiveness of coagulation over PAC for the removal of the high molecular weight compounds. These observations are consistent with past studies.[2-4]

The very high turbidity removal achieved with UF membranes (Table 2) made sense after we measured the particle size distribution of the reconstituted Lake Decatur water with the electrical sensing zone technique (Figure 4). Note that the vast majority of particulate mass (volume) has a size above the minimum resolution of the counter, about 1.5 μm. Since typical nominal pore sizes for the membranes used in this work are on the order of 0.01 to 0.02 μm,[5] the efficiency of UF membranes in removing turbidity is understandable.

Before proceeding with the batch and semicontinuous experiments using the more expensive hollow fiber membranes, we did a considerable amount of pretreatment screening using Amicon disk membranes and the "dead-end" batch filtration cell shown in Figure 1. Flux measured during a filtration test is designated J. The initial flux through the clean membrane (also called the initial permeability) is designated J_0. The membrane final permeability is called J_f. The

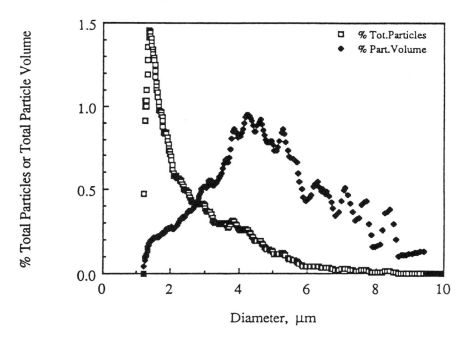

FIGURE 4. Particle size distribution of reconstituted Decatur water (02-13-89).

units of all these flux measures are liters per (square meter surface area × hour × atmosphere of pressure). Figure 5 shows some flux results for the XM- and YM100 membranes using different pretreatment steps and combinations. First notice that the more hydrophilic YM-series membrane is considerably less susceptible to irreversible fouling: J/J_o and J_r/J_o are both consistently higher for the YM-series membranes, regardless of the pretreatment. Pretreatment is of obvious importance in maintaining flux, especially for the YM-series membrane: coagulation or coagulation plus PAC pretreatment increases J/J_o from around 20% to the range of 60 to 80%. For the relatively hydrophilic YM-series membrane, we believe that the main mechanism for this increased flux is the formation of a more porous cake on the membrane surface. Adsorptive fouling is relatively insignificant for this membrane material. For the relatively hydrophobic XM-series membrane, improvements in flux and flux restoration with pretreatment are related both to the removal of organic adsorbents and the porosity of the surface cake (see later modeling discussion).

TOC removal efficiency varies more with pretreatment than with membrane polymer type (Table 2, Figure 6). Figure 7 compares pretreatment methods for the removal of TOC and THMFP using the YM-series membrane. Simple

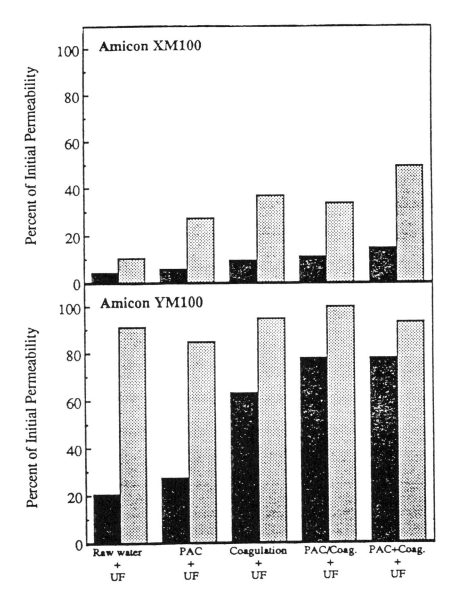

FIGURE 5. Effect of different treatment processes on flux and final permeability for XM50, XM100, and YM100 membranes.

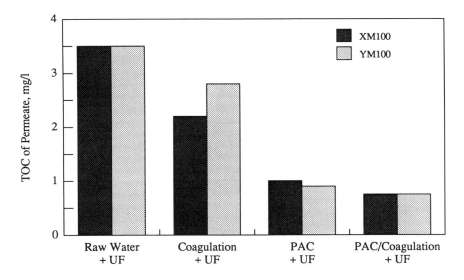

FIGURE 6. Effect of pretreatment on the TOC after ultrafiltration (initial TOC = 6.1 mg/L).

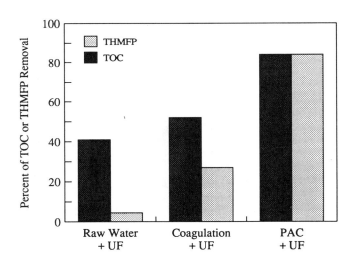

FIGURE 7. Comparison of different UF processes for TOC and THMFP removal using YM100 membrane.

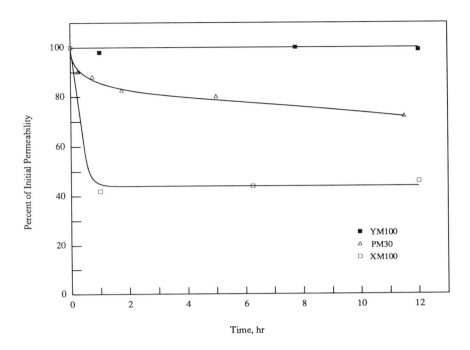

FIGURE 8. Comparison of flux after static adsorption for YM, PM, and XM membranes. (Reprinted from *J. AWWA*, Vol. 81, No. 11 [November 1989]. By permission of American Water Works Association.)

membrane filtration of the lake water does not lead to a very high reduction in THMFP: apparently the most reactive organic fraction is the small molecular weight TOC fraction isolated in the gel permeation chromatography results. Not until this small molecular weight fraction is complexed and associated with a particulate phase during pretreatment (coagulation and/or PAC) can we achieve important THMFP removal with UF. Using PAC we can achieve THMFP removals in the range of 80 to 85%.

Since we believed that fouling of our membranes was related to adsorption of organic compounds on the membrane, we developed static membrane adsorption tests to test our hypothesis. If significant pore fouling was occurring due to adsorption of organic material, then one would expect the membrane permeability to change. Figure 8 shows results for the three disk membranes we used. Our hypothesis about the cause of fouling seems to be supported with these results: the PM- and XM-series membranes, which showed the most irreversible fouling in filtration tests, also showed a high amount of pure adsorptive flux loss in the static adsorption tests. The relatively hydrophilic YM-series membrane is relatively unaffected by adsorptive fouling.

Table 3. Series Resistance Calculations for U.F. Membranes (Resistance Expressed in Units of $(hr/m^2/atm^2/L) \times 10^{-2}$: % R as Percent of Total Resistance)

Membrane	R_M	R_c	R_i	R_T	% R_M	% R_c	% R_i
YM 5	15.05	0.19	0.65	15.89	94.71	1.19	4.10
YM 100	0.74	2.78	0.05	3.57	20.73	77.87	1.40
XM 100	0.35	5.02	2.92	8.29	4.22	60.55	35.23
PM 30	0.38	1.85	0.44	2.67	14.23	69.29	16.48

Reprinted from *J. AWWA*, Vol. 81, No. 11 (November 1989). By permission of American Water Works Association.

The series resistance model has been proposed by several investigators to model flux in membranes. The model sometimes takes the following form:

$$J = \frac{\delta P}{R_m + R_c + R_i}$$

Here δP is the transmembrane pressure drop, R_m is the hydraulic resistance of the membrane, R_c is the resistance of the cake layer, and R_i is the resistance due to a flux loss from some irreversible adsorption. Using this model and our initial, final, and restored flux values, we have calculated the various resistance terms (Table 3).

One might hypothesize that although a cake layer forms on both the YM-5 and YM-100 membranes (as observed), it does not significantly affect flux of the YM5 because R_m is much greater than either R_c or R_i. Referring to Table 3, our hypothesis seems consistent. For the YM-5 membrane, the cake resistance is only estimated to be about 1% of the total flux resistance; for the YM-100 membrane, the cake resistance is calculated to be almost 78% of the total flux resistance. The easy removal of the cake layer explains the large reversible flux loss component. We also noted here that irreversible flux losses are much more severe for the XM- and PM- series membranes, and while hydraulic losses due to cake formation can be significant for the YM-100 membrane, these losses are by and large reversible. The series resistance calculations in Table 3 are consistent with these observations. The high degree of flux restoration available with the YM-series membranes makes this material an interesting candidate for use with natural waters.

Two Amicon hollow fiber modules, the H1P100 (100-K MWCO) and the H1P10 (10-K MWCO), were used in the pilot shown in Figure 2. The H1P-series membranes are made of polysulfone. Flux and flux recovery for the hollow fiber modules operated in the continuously circulated batch mode are shown in Figure 9. First note the severe fouling for filtration of the unpretreated Lake Decatur water (J/J_0). Second note that while the addition of coagulation pretreatment does little to enhance flux for the 100-K MWCO membrane, it more than doubles the ending flux values for the 10-K MWCO membrane. More

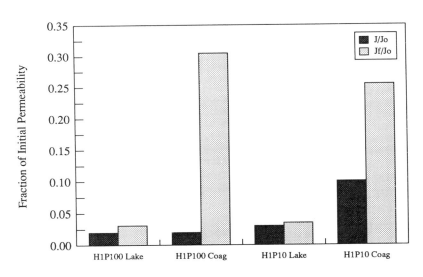

FIGURE 9. Flux and flux recovery for H1P100 and H1P10 membranes treating Lake Decatur and coagulated Lake Decatur water.

importantly, coagulation pretreatment significantly increases recovery after backflush (J_f/J_o).

A spaghetti or straw-like sludge layer was ejected from the hollow fibers during backflushing with clean water. The diameter of this sludge material is exactly the inside diameter of the hollow fibers (1.1 mm): apparently, this material is a compact layer of coagulated lake water solids formed along the inside of the fibers. We felt that the easy removal of this layer from the hollow fibers shows that rather than plug the fibers, coagulation pretreatment results in the formation of a dense "filter cake" which could obviously be processed fairly easily. The residual from UF treatment of a coagulated surface water is less of a "brine" and more of a conventional water treatment sludge.

In the semicontinuous operation mode, a 20-L sample (treated or untreated) is circulated through the hollow fiber system and all permeate and retentate are returned to the 20-L vessel. After about 400 to 500 L of permeate passes through the membrane (this takes about 3 hr), the system is turned off, the membrane is backflushed, a new sample is added to the vessel, the backflushed membrane is replaced in the pilot, and filtration is resumed with the new solution. Figure 10 shows typical results from this semicontinuous operation, where we see flux decline and recovery after backflush and sample replacement. Figure 11 shows the permeate TOC and turbidity during treatment of coagulated water using the backflush/sample replacement operation mode. Interestingly, there may be some kind of conditioning of the membrane over time, as there seems to be a steady

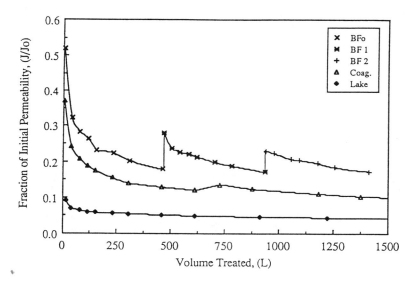

FIGURE 10. Flux for the H1P10 membrane treating reconstituted, coagulated, and coagulated with intermittent backflushing Lake Decatur water: J_o (lake) = 768, J_o (coagulated) = 458, J_o (BF) = 276 L/hr m² atm.

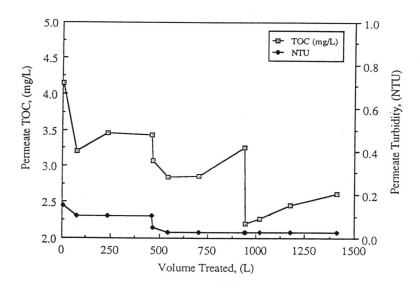

FIGURE 11. Total organic carbon and turbidity analysis on permeate for H1P10 membrane treating Lake Decatur water with coagulation/intermittent backflushing and sample replacement after backflush.

improvement in TOC and turbidity removal over time and as the number of backflushes and sample replacements continues.

CONCLUSIONS

Even without pretreatment, ultrafiltration can reliably decrease the lake water turbidity to a level less than 0.1 NTU. For the Lake Decatur water with no pretreatment, most organic matter (~60%) will pass through typical UF membranes. Irreversible flux loss seems to be due to adsorption of organic compounds on the membrane surface and/or in the pores. Although cake layers do build up on the membrane surface, they are generally removable with surface and backflushes. Permeate flux is in the pressure-controlled region (i.e., no mass-transfer limitation).

With or without pretreatment, one of the most important considerations in the filtration of natural waters is the membrane material. Hydrophilic membranes (e.g., cellulose acetate) seem to suffer much less adsorptive flux loss. Unfortunately, the most commonly available UF membranes are polysulfone, which are relatively hydrophobic. This is because polysulfone membranes are more resistant to chlorine and extreme pH-values (important qualities in industries like food processing).

Ultrafiltration without pretreatment is not much better than conventional processes for removing TOC and is probably worse than conventional processes for removal of THMFP. However, UF with coagulation pretreatment can significantly increase THMFP removal. Powdered activated carbon pretreatment with or without coagulation can lead to TOC and THMFP removals in the range of 80 to 85%. Coagulation pretreatment does significantly increase membrane flux and decrease the amount of irreversible membrane fouling. Therefore, a pretreatment scheme using coagulation and PAC could achieve the dual goals of high flux and high TOC/THMFP removals.

To be competitive with existing technologies for the treatment of typical midwestern (U.S.) surface waters, future UF research should concentrate on optimization of pretreatment for THMFP removal and high flux values, and development of membrane cleaning techniques for combatting irreversible membrane fouling.

ACKNOWLEDGMENTS

The author thanks the Lyonnaise des Eaux/Dumez, Paris, France, for supporting most of the work reported here. Some of the work reported here was funded through the U.S. Environmental Protection Agency under assistance agreement EPA Cooperative Agreement EPA CR 812582 to the Advanced

Environmental Control Technology Research Center. This work is not subject to the Agency's required peer and administrative review and therefore does not necessarily reflect the views of the Agency and no official endorsement should be inferred. Mention of trade names of commercial products does not constitute endorsement or recommendation for use.

This paper was based on the Master's Theses of Jean-Michel Laîné (1989) and James P. Hagstrom (1989).

REFERENCES

1. Taylor, J. S., Thompson, D. M., and Carswell, J. K. "Applying Membrane Processes to Groundwater Sources for Trihalomethane Precursor Control," *J. AWWA* (August 1987), pp. 72-82.
2. Schnoor, J. L., Nitzschke, J. L., Lucas, R. D., and Veenstra, J. N. "Trihalomethane Yields as a Function of Precursor Molecular Weight," *Environ. Sci. Technol.* 13(9):1134-1138 (1979).
3. Collins, M. R., Amy, G. L., and Steelink, C. "Molecular Weight Distribution, Carboxylic Acidity, and Humic Substances Content of Aquatic Organic Matter: Implication for Removal During Water Treatment," *Environ. Sci. Technol.* 20(10):1028-1032 (1986).
4. Randtke, S. J. "Organic Contaminant Removal by Coagulation and Related Process Combinations," *J. AWWA* (May 1988), pp. 40-56.
5. Fane, A. G., Fell, C. J. D., and Waters, A. G. "The Relationship Between Membrane Surface, Pore Characteristics, and Flux for Ultrafiltration Membranes," *J. Membr. Sci.* 9:245-262 (1981).
6. Laîné, J.-M. "Optimization of Organic Removal in the Ultrafiltration of a Natural Water," Master's Thesis, Department of Civil Engineering, University of Illinois, Urbana-Champaign, IL (1989).
7. Hagstrom, J. P. "Performance of Polysulfone Membranes in the Ultrafiltration of a Surface Water," Master's Thesis, Department of Civil Engineering, University of Illinois, Urbana-Champaign, IL (1989).
8. Amicon, Inc. Product catalog (1987).
9. Lee, C. K., and Hong, J. "Characteristics of Electric Charges in Microporous Membranes," *J. Membr. Sci.* 39:79-88 (1988).
10. Cheryan, M. *The Ultrafiltration Handbook* (Lancaster, PA: Technomic Publishing Co., Inc., 1988).

CHAPTER 24

Megatrends in Drinking Water Treatment Technologies: The Years Ahead 1990

F. Fiessinger

INTRODUCTION

In early 1985, I wrote with a group of researchers of the Lyonnaise des Eaux Research Center, a general paper[1] on where we felt that the technologies of drinking water treatment were heading. This paper was first presented at a workshop on "Environmental Technology Assessment" held at the University of Cambridge in England on April 24, 1985. It was then reproduced at the end of 1985 in *The Water Research Quarterly* of the AWWA Research Foundation, and it got through this unexpected publicity throughout the United States. The title of the paper, but also its content, had been considerably influenced by the famous book and #1 bestseller, *Megatrends*.[2] Almost 5 years later I am asked to make a revised version of this article. Following the analogy with John Naisbitt, I have called it: "The Years Ahead 1990."[3] It does not pretend to be a comprehensive review of all foreseeable evolutions of water treatment technology, but rather focuses on the few trends which appear to me today, the most capable of bringing appreciable changes in drinking water treatment practices in the course of the next decade.

THE PRESSURES TO CHANGE

All around the world the water supply profession is undergoing accelerated changes, which usually translate into an increase in research efforts. These changes are due in part to the strategy of the water service itself. They are, however, primarily created by external forces beyond the control of the water supply profession. Ten of these exogenous forces can be easily identified.[4] Their importance varies considerably from one country and even one region to another. The list is not comprehensive and each of them may interact with the others.

1. The tightening of quality standards accompanied by an overall progress in consumer demands.
2. The progress in technology, such as electronics, biotechnology, polymer science, membrane separation, etc., which means an increase in the general level of knowledge and exerts pressure on all industrial sectors to introduce changes.
3. The spinoff from the development of treatment of water for specific applications, such as ultrapure water for laboratories, water for the electronic industry, for enhanced oil recoveries, bottled water, etc.
4. The quantitative and qualitative reduction in water resources. This is sometimes related to pressure for a better environment (e.g., the state of California).
5. The need to recycle water (e.g., in the buildings in Tokyo).
6. The economic pressure exerted to reduce energy and labor costs, and increase productivity.
7. The development of privatization resulting in competition between water organizations. This entails the need for greater management efficiency and brings somewhat into question direct local public management. It also forces public services to look for new markets and therefore develop a policy of diversification. The most common one is water treatment.
8. Reduction in the market for new construction works, notably because of financing difficulties on international markets.
9. Network deterioration, maintenance, and renewal problems. The specific problem of the uncertainty of the life span of networks may weigh heavily on decisions for investing in new water treatment plants. This, combined with the need for a better quality water, reopens the door to dual distribution systems.
10. Changes in the mentality of water works personnel and "detailorization" of tasks.

Today, more than ever it seems that in this list the strongest trends enhancing changes in water treatment are the following:

- The need for quality and the new U.S. Safe Drinking Water Act (1)
- The progress in technology and in turn the demand for "high-tech"[2]
- The development of privatization,[3] more particularly in Europe, where the completion of the single market after 1992 might accelerate changes.

The conjunction of several of these forces might literally create "windows of opportunity" for specific technologies in the coming few years, and it is likely in fact that most of the changes will occur in the early part of the 1990s.

THE RESEARCH ORIENTATIONS

As indicated earlier, the rise in R&D expenditures and the increasing involvement of water utilities are a good sign of the changes to come. In these expenditures, R&D in water treatment usually represents more than one third. It is thus expected that the changes will occur here at a faster rate than in water distribution, metering, pipe laying, or consumer management.

What kind of research on water treatment is being done? In general this R&D seems to be rather conservative: it focuses more on a better understanding of the internal mechanism and a better control of existing processes than on the development of new ones. A large portion of the effort is also spent in pilot experiments, which are usually local demonstrations, of a proven technology.

As an example among the papers selected in the journal *AWWA* for their 1989 best paper award:

- 30% were on activated carbon absorption
- 25% were on ozone oxidation
- 10% were on filtration
- 10% were on membrane separation
- 10% were on air stripping
- 10% were on sludge dewatering

Carbon absorption and ozone oxidation still represent more than half of the research efforts. This is a rather general situation, but unlike 5 years ago, new technologies such as membrane separation are now appearing in the programs and some research organizations are deliberately putting the majority of their strengths on innovative technologies. I believe now much more than 5 years ago that we have here the seeds for a technological revolution.

A FOCUS ON ORGANIC MATTER REMOVAL

It is needless to say, at a workshop on "Influence and Removal of Organics", that water treatment will, in the years to come, continue to focus on an improvement of the removal of the organic constituents. The push for reinforced standards following the U.S. Safe Drinking Water Act, the need to eliminate all the new synthetic organics appearing in the environment, the need to avoid the formation of disinfection by-products, the need to control the bacterial regrowth in the distribution systems, etc., all these reasons will lead to technologies more and more effective in organics removal. This does not mean of course that local problems caused by other substances may not arise. Compounds such as Radon, heavy metals, radioactive metals, barium, and of course calcium, sodium, potassium, etc. will always generate treatment technology improvements. The main thrust will, however, remain on the organics for the years to come.

Together with treatment technologies, tremendous efforts and progress will be made in the areas of water analysis, identification of the mechanisms of formation of the by-products, and health effects assessment; and the next decade might see the development of toxicity bioassays.

ABSORPTION, OXIDATION, AND DISINFECTION

These treatment technologies which are traditionally being used for the reduction of organic matter will naturally continue to grow in importance. A better usage of granular activated carbon (GAC) through a better modeling of the absorption mechanisms seems to be the surest evolution of current practice. Alternatives to GAC such as activated alumina, which raised some interest in the mid-80s is now fading out. New carbons, with various shapes (fibers?), might appear, better reactivation technologies will develop, but altogether this will result only in minor changes over the technologies which have been in use for the past 15 years. Powdered activated carbon (PAC) will probably regain importance over GAC, particularly in conjunction with floc blanket clarifiers and new membrane reactors (see later).

The development of ozone will continue. It will be more and more systematically applied together with GAC, but this again is an old trend. The technology of its generation and of its contacting with water will consistently be improved. The newest trend is probably in the development of the combination of ozone with other oxidants such as UV and H_2O_2. New reactors allowing an efficient combination of these oxidants will naturally be developed.

Chlorine use will remain as a safety disinfectant, at very low dosages, after removal of the core of the organics and of all potential precursors. Chlorine dioxide will probably remain of limited usage and will not undergo any significant development over the existing situation.

BIOLOGICAL TREATMENT

It is still hoped that future water treatment technologies will make a greater use of biological processes. Nitrification and more particularly denitrification processes will be developed. It is likely that some form of biological treatment will be systematically used at the water plant for a better control of the bacterial regrowth in the distribution system. This is usually combined with GAC absorption. The removal of trace organics by the nitrifying biofilm, the development of biofilters for metal precipitation, etc. will all undergo interesting developments.

The control of the biofilm in the distribution system, where disinfectants will be at trace levels, will be of major concern. No significant improvement, however, seems likely to occur and although the steady rise in research efforts in this area will continue, the traditional weaknesses of biological treatments i.e., lack of reliability and strong sensitivity to flow and loadings changes, will cripple the expansion of their application to drinking water production.

CLARIFICATION

Five years ago, I expected an important change in this area due in particular to the arrival of the polymeric coagulants. It did not occur and it is more and more unlikely that it will! The quest for quality renders more and more difficult the use of any chemical. The replacement of conventional inorganic coagulants such as alum and ferric chloride by more efficient polymeric species is progressing very slowly. In parallel, rising suspicions on the usual effect of aluminum residual in Alzheimer's disease may definitely condemn the development of this metallic coagulant.

Clarifiers and sand filters technology seems to have reached a plateau and no new development is likely to occur. Much more research work will be done, however, in this area because of the new U.S. regulations on the filtration requirement and the guideline turbidity (0.5 JTU). This regulation will undoubtedly open a window of opportunity for membrane filtration. If membranes do not appear to be effective technologies for turbidity removal, by the end of 1991, conventional sand filters may be built and membranes may wait many more years for their industrial application.

MEMBRANE SEPARATION

Among all the water treatment technologies, membranes seem to be the most promising ones. Their best advantage is their ability to produce water with a constant and very well-adjusted quality. At a time where the demand for quality is a major drive, membranes thus have a clear advantage over all other technologies. In addition, membranes offer a variety of other advantages. They remove a wide range of substances, from particles to ions, including bacteria and viruses. Theoretically they could remove everything. They can operate without chemical addition to water, are reliable, compact, and easy to automate.

Their major disadvantages are still their rather high capital and operating costs, and the fact that they are prone to fouling, which requires often a high level of pretreatment and regular chemical cleaning. They may also have a rather important reject stream whose disposal may create problems.

Membranes can be made in a wide variety of shapes: fibers, tubes, flat sheets in the form of spiral wound modules or plate and frames, etc. They can have an almost unlimited range of porosity — or molecular weight cutoffs — and the development of synthetic organic chemistry gives hopes of great improvement in the membrane composition. This will certainly result in lower operating pressures, decreased fouling, better resistance to disinfectants, biodegradation, etc. and altogether a much longer lasting period. The new composite membranes with sophisticated coatings have now little to do with the original cellulose acetate membranes and the progress will likely accelerate in the coming few years.

Several outside forces are pulling the membrane technology toward drinking water application. The first and primary force is the development of a huge market in Florida where plants with a total capacity of roughly 200 MGD are already in operation and where at least an additional 100 MGD will be built every year. Membranes in Florida are not only used for desalination, but more and more for organics removal (THMFPs) together with softening. Looser membranes as opposed to conventional RO —in the range of nanofiltration and ultrafiltration — are thus being installed and make the treatment more and more cost effective as compared to conventional lime precipitation. Total costs for a plant of 0.5 MGD with membranes are in the range of 1.5 $/1000 gallons as compared to more than 2. $/1000 gallons for lime softening. This difference decreases, however, when the size of the plant increases. Membranes are now moving northward and westward throughout the United States, but Florida will remain indeed the nest for membrane proliferation in the drinking water sector.

Another reason for the progress of membrane technology is the development of considerable international research programs such as the Aquarenaissance program in Japan and the Eureka program in Europe. The latter in particular fuels the development of ultrafiltration and microfiltration membranes at Lyonnaise des Eaux/Dumez, which are capable of meeting the costs of conventional filtration technologies with a whole load of technical advantages. Plants are already in operation and if these membranes can take advantage of the window of opportunity opened by the SDWA in the U.S., they will bring one of the most important revolutions in water treatment practices. In parallel the Eureka program allows the Canadian firm Zenon to develop among others highly specific membrane fibers for the removal of color, which is a common problem in all northern countries.

An interesting feature of membranes is their ability to constitute a reaction by combining the separation with a reaction in the recirculation loop. A whole world of membrane reactors can thus be envisioned: absorption reactors with powdered activated carbon in MF membranes, oxidation reactors, carbonate precipitation, and iron precipitation reactors; the best hope being in the membrane/bioreactors, which would significantly open the door to industrial applications of biotechnology to water treatment.

A last point of potential development is the system (Figure 1) where membranes are included. Membranes are too often studied independently. They are an

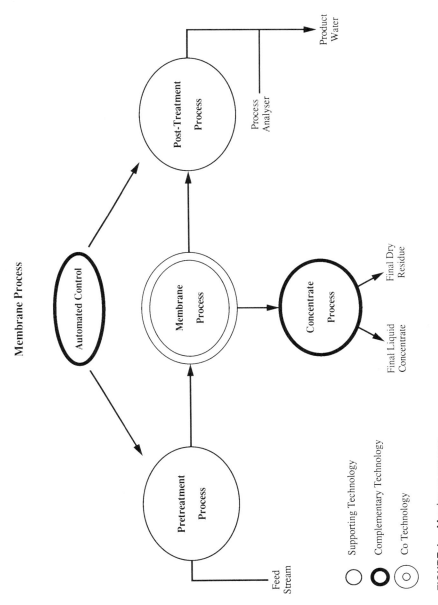

FIGURE 1. Membrane process.

essential, but only small component of an overall system which leaves considerable room for improvement. The "system" approach will likely result in great progress.

AUTOMATION

Our 1985 article put great hopes in the development of new computer technologies. Five years later we realize that it is coming slower than expected, but the hopes are still here. In fact I believe that not only artificial intelligence, but also other forms of computer architecture such as in particular neural networks will bring tremendous changes to all forms of water treatment technologies. They will constitute powerful control tools capable of handling at a fantastic speed fuzzy data and more than anything capable of automatic learning. They may bring back to life obsolete technologies and will drastically boost the development of the most sophisticated ones. It seems that all water plants will be equipped in the course of the next decade with some form of these knowledge based systems.

CONCLUSIONS

Looking back at our 1985 predictions we can proudly acknowledge that they do not seem too wrong. We underestimated the potential of membranes and we expected a faster speed of change. Change has now started and on it goes exponentially. Five years from now we might realize that things had gone much faster than I could predict today. In any case it would certainly be the best thing which could happen to the quality of the water supplied around the world and, in turn, to us all.

REFERENCES

1. Fiessinger, F., J. Mallevialle, A. Leprince, and M. Wiesner. "Megatrends in water treatment technologies," *Water Res. Q.*, AWWA Research Foundation, Denver, CO, January 1986.
2. Naisbitt, J. *Megatrends* (Warner Books, 1984).
3. Naisbitt, J. et al. *The Years Ahead 1986* (Warner Books, 1985).
4. Fiessinger, F. "Applied Research in Water Supply," IWSA, Rio de Janeiro, September 1988.

Contributors

Jeffrey Q. Adams, Drinking Water Research Division, U.S. Environmental Protection Agency, 26 W. Martin Luther King Drive, Cincinnati, Ohio 45268

C. Anselme, Lyonnaise des Eaux/Dumez, Central Laboratory, 38 Rue du President Wilson, F-78230 Le Pecq, France

P. Aptel, Lyonnaise des Eaux/Dumez, Central Laboratory, 38 Rue du President Wilson, F-78230 Le Pecq, France

R. Audebert, University Pierre, Marie Curie, Laboratory of Macromolecular Phys/Chem, 10 Rue Vauquelin, F-75231 Paris, Cedex 05, France

J. L. Bersillon, Lyonnaise des Eaux/Dumez, Central Laboratory, 38 Rue du President Wilson, F-78230 Le Pecq, France

Lin Bingjie, Harbin Water Supply Company, Harbin, China

Diane J. W. Blum, Ph.D., P.E., Blum Environmental Engineering and Research, 112 Clwyd Road, Bala Cynwyd, Pennsylvania 19004

Jean-Yves Bottero, Centre National de la Recherche Scientifique, Laboratoire Environnement et Mineralurgie, Rue du Doyen Marcel Roubault, F-54501 Vandoeuvre, France

E. Brodard, Lyonnaise des Eaux/Dumez, Laboratoire de Recherche, 38 Rue du President Wilson, F-78230 Le Pecq, France

Auguste Bruchet, Lyonnaise des Eaux/Dumez, Central Laboratory, 38 Rue du President Wilson, F-78230 Le Pecq, France

Jean Maurice Cases, Centre National de la Recherche Scientifique, Laboratoire Environnement et Mineralurgie, Rue du Doyen Marcel Roubault, F-54501 Vandoeuvre, France

Mark M. Clark, University of Illinois, Department of Civil Engineering, 205 North Mathews Avenue, Urbana, Illinois 61801

C. Corless, Imperial College, Department of Civil Engineering, London SW7 2AZ, England

Wang Dazhi, Harbin Water Supply Company, Harbin, People's Republic of China

Jean Pierre Duguet, Lyonnaise des Eaux/Dumez, Central Laboratory, 38 Rue du President Wilson, F-78230 Le Pecq, France

Graziella Durand, University Pierre, Marie Curie, Laboratory of Macromolecular Phys/Chem, 10 Rue Vauquelin, F-75231 Paris, Cedex 05, France

Jun-Sheng Feng, The Macao Water Supply Company, Ltd., 82 Avenida do Conselheiro Borja, Macau

Yubi Feng, The Macao Water Supply Company, Ltd., 82 Avenida do Conselheiro Borja, Macau

Francois Fiessinger, SITA, 7 rue de Logelbach, F-75017 Paris, France

Nigel J. D. Graham, Imperial College of Science and Technology, Senior Lecturer in Environmental Engineering, Department of Civil Engineering, London SW7 2BU, England

Thomas J. Haley, National Center for Toxicological Research, Scientific Intelligence, Jefferson, Arkansas 72079

Zhang Hua, Shanghai Municipal Engineer Design Institute, Scientific Research Department, 133 Yuan Ming Yuan Road, Shanghai 200002, China

Conrad Hubele, Philipp Muller GmbH, Monchstrasse 11, 7000 Stuttgart 10, Germany

Zhanpeng Jiang, Tsinghua University, Department of Environmental Engineering, Beijing 100084, People's Republic of China

Xu Jianhua, Tongji University, School of Environmental Engineering, 1239 Siping Road, Shanghai 200092, People's Republic of China

Fan Jinchu, Tongji University, School of Environmental Engineering, 1239 Siping Road, Shanghai 200092, People's Republic of China

T. Kamei, Hokkaido University, Department of Sanitary and Environmental Engineering, Nishi 5, Kita 8, Kita-Ko, Sapporo 060, Japan

Francoise Lafuma, University Pierre, Marie Curie, Laboratory Macromolecular Phys/Chem, 10 Rue Vauquelin, F-75231 Paris, Cedex 05, France

Benjamin W. Lykins, Jr., Drinking Water Research Division, U.S. Environmental Protection Agency, 26˙W. Martin Luther King Drive, Cincinnati, Ohio 45268

Joël Mallevialle, Lyonnaise des Eaux/Dumez, Central Laboratory, 38 Rue du President Wilson, F-78230 Le Pecq, France

Jacques Manem, Lyonnaise des Eaux/Dumez, Central Laboratory, 38 Rue du President Wilson, F-78230 Le Pecq, France

Issam N. Najm, James M. Montgomery, Consulting Engineerings, Inc., Applied Research Department, 250 N. Madison Avenue, Pasadena, California 91101

CONTRIBUTORS

Roger Perry, Imperial College, Department of Civil Engineering, London SW7 2AZ, England

G. Reynolds, Imperial College, Department of Civil Engineering, London SW7 2AZ, England

Yves Richard, DEGREMONT, CIRSEE, 38 Rue du President Wilson, F-78230 Le Pecq, France

Bruce E. Rittmann, University of Illinois, 3221 Newmark Laboratory, Department of Civil Engineering, 205 North Mathews Avenue, Urbana, Illinois 61801

M. Roustan, I.N.S.A, Department Genie des Procedes Indus., F-31077 Toulouse Cedex, France

Zhou Shaojia, Guangzhou Water Supply Company, Guangzhou, China

Vernon L. Snoeyink, University of Illinois, Department of Environmental Engineering, Tsinghua University, Beijing 100084, People's Republic of China

I. H. Suffet, Drexel University, Environmental Studies Institute, Philadelphia, Pennsylvania 19104

N. Tambo, Hokkaido University, Faculty of Engineering, Department of Sanitary and Environmental Engineering, North 13, West 8, Sapporo 065, Japan

Tong Kuan Wang, Laboratoire de Physico-Chimie Macromoleculaire, Universite Pierre et Marie Curie, Unite Associee au CNRS n 278, ESPCI, 10 Rue Vauquelin, F-75231 Paris, Cedex 05, France

Z. S. Wang, Tsinghua University, Department of Environmental Engineering, Beijing 100084, China

Chen Wanhua, Guangzhou Water Supply Company, Guangzhou, China

Mark R. Wiesner, Rice University, Environmental Science and Engineering, P.O. Box 1892, Houston, Texas 77251

Yuet-Nar Wong, The Macao Water Supply Company, Ltd., 82 Avenida do Conselheiro Borja, Macau

J. X. Yang, Tsinghua University, Department of Environmental Engineering, Beijing 100084, China

Z. H. Yang, Tsinghua University, Department of Environmental Engineering, Beijing 100084, China

Akira Yuasa, Gifu University, Faculty of Engineering, Department of Civil Engineering, 1-1, Yanagido, Gifu-city 501-11, Japan

Ye Zheng-Zhong, Shanghai Municipal Engineering Design Institute, 133 Yuan Ming Yuan Road, Shanghai, People's Republic of China

Li Zhicai, Guangzhou Water Supply Company, Guangzhou, People's Republic of China

W. P. Zhu, Tsinghua University, Department of Environmental Engineering, Beijing 100084, China

Index

acclimation, 289, 297
acetone, 74
acetophenone, 74
acrolein, 75
acrylic copolymer membrane, 314
activated carbon, 187, 191
  adsorption of organic chemicals on, 67–78
    comparison of four SAR methods, 70–73
    data, 68–70
    LSER correlation for expanded database, 73–76
  competitive adsorption on, 19
  granular, 35, 79, 207–217
activated carbon in water, competitive adsorption of several organics and heavy metals on, 79–95
  experimental procedures, 79–81
    analytical methods, 80
    experimental method, 80–81
    materials, 79–80
  results and discussion, 81–93
    bisolute systems, 84–90
      humic acid-Cd(II), 87–88
      humic acid-Cr(VI), 88–90
      phenol-Cd(II), 86–87
      phenol-Cr(VI), 87
      phenol-Pb(II), 84–86
    single solute systems, 81–84
      Cd(II), 83–84
      Cr(IV), 84
      humic acid, 81–82
      Pb(II), 83
      phenol, 81
    X-ray photoelectron spectrum, 91–93
adsorbed phase, 23
adsorption capacity, variation of, 85
adsorption isotherms, 5, 8, 117
adsorption kinetics, 38
adsorptive fouling, 313
air stripping, 327
alcohols, 69, 72
aldehyde(s), 69, 258

adsorption correlated for, 72
derivatives concentration, 247
standards, 256
aliphatic chemicals, 70
aliphatic hydrocarbons, 251
alkanes, 251, 253, 255, 258
alkenes, 251, 253, 255
Al NMR patterns, solid and liquid state, 104
aluminum hydroxide polycations, 97
Ames test, 229–231
4-aminobiphenyl, 74
ammonia, removal of, 228–229
ammonia-nitrogen, bionitrification of, 280
anthracene, 74
aromatic hydrocarbons, 251, 253
aromatics, 262
astringent sensation, 247
atrazine, removal by conventional treatment processes, 50

background organics, 20
  correlation of p-aminobenzoic acid and, 32
  isotherm of, 28, 31
batch adsorption equilibria, simulations of, 22
batch adsorption isotherm, 19
batch isotherms, shapes of, 23
batch tests, 24
batch ultrafiltration, 312
bed fluidization, 279
benzene, removal of by conventional treatment processes, 50
benzo(a)pyrene, 74
benzo(b)fluoranthene, 74
benzo(ghi)perylene, 74
benzoic acid, 74
benzo(k)fluoranthene, 74
biocides, 48
biodegradable organic matter (BOM), 291
biodegradable oxidation by-products, 267
biodegradation, 290
biodegradation kinetics, 295
biofilm, 289, 290, 329

337

biological filtration, removal of synthetic and natural organic compounds by, 289–298
  removal of natural organic polymers, 290–294
  removal of synthetic organic compounds, 294–296
biologically stable water, 289
biological oxidation, 285
bisolute system
  adsorption capacity in, 85
  adsorption isotherms at different pHs, 88–90
  adsorption isotherms of phenol and Cd(II) in, 86–87
BOM, see biodegradable organic matter
bottle-point technique, 42
Bourbince river, water treatment, 166–170
Bragg-Williams approximation, 4
1-butanol, 74
2-butanone, 74
butyraldehyde, 75

capyl pyridinium chloride (CPC), 257, 259
carbohydrates, 290
carbon tetrachloride, 50, 74
carboxylic acids, 259
cationic polymer(s), 116
cationic surfactants, 257
Cd(II), adsorption capacity of, 83–84, 87
charge neutralization, mechanism due to, 109
China, water treatment in, 137–141, 187–193
China National Health Standard for Drinking Water, 140
chlorinated benzenes, 295
chlorinated phenols, 295
chlorine, 55, 252, 328
chlorobenzene, 74
chloro-benzenic compound, 267
chloroethanes, 51
chloroform, 74
  mean content in treated and raw water, 140
  removal of, 52, 226
chlorohydrocarbons, 251
2-chlorophenol, 74
clarification, 157, 300
clarifiers, 329
clay, 116
closed-loop stripping analysis (CLSA) method, 208, 242
CLSA-GC/MS, 214
coagulant doses, 174

coagulants, 300
coagulation-flocculation, of minerals using Al,Fe(III) salts, 97–114
  discussion, 108–112
  experiments, 99–103
    flocculation experiments, 99–101
    small-angle X-ray scattering, 101–103
    materials, 99
  results, 104–107
    chemistry of flocculants at neutral pH, 104–105
    flocculation and electrophoresis results, 105–107
    small-angle X-ray scattering, 107
coagulation mechanisms, 115
coagulation pretreatment, 320, 321, 323
coagulation process, treatability of, 144
coagulation-sedimentation-filtration system, 221
colloid, 125, 300
color, removal of, 226, 330
color-producing molecules, removal of, 79
column adsorber, 21
column isotherm, 22
column tests, 24
contaminated raw water source, fluidized biofilter bed process as preliminary treatment of, 279–287
  basic principle, 280
  characteristics of FBBP, 285–286
    detachment of biofilms, 286
    fluidization, 286
    supply of oxygen, 285–286
  efficiency of FBBP, 281–282
  experimental equipment, 281
  experimental study of FBBP, 281
  investigation of general characteristics of FBBP, 285
  normal operational stage, 283–284
continuous-flow slurry contactor, 39
converged isotherm, 24
copolymers, 116
corrosion, 291
CPC, see capyl pyridinium chloride
Cr, adsorption capacity of, 84
Cr(VI), adsorption capacity of, 90
critical micellar concentration, 1, 6
CT concept, 195
cut-off threshold, 300
cyclohexanone, 75

2,4-D, 48
decanoic acid, 256
2-decanone, 258

INDEX

dec-1-ene, 255
Deep U Tube, 196
denitrification, 289, 290, 328
depth filtration, 125
desalination, 330
detergent removal, 40
1,2-dichlorobenzene, 74
1,3-dichlorobenzene, 74
1,4-dichlorobenzene, 74
dichloroethane, removal of by conventional treatment processes, 50
1,1-dichloroethane, 74
1,2-dichloroethane, 74
dichloroethenes, 51
1,1-dichloroethylene, 74
trans-1,2-dichloroethylene, 74
dichloromethane, 138
2,4-dichlorophenol, 74
1,2-dichloropropane, 74
dimethyl dialkyl ammonium chloride, 258
2,4-dimethylphenol, 74
di-n-butylamine, 75
2,6-dinitrotoluene, 74
di-n-propylamine, 75
di-n-propyl ether, 75
1,3-dipolar cycloaddition, 258, 259
direct filtration, 300
disinfection, 187, 192, 193
distance distribution functions P(r), 107
docos-1-ene, 255
dodecane, 255
dodecanoic acid, 256
dodec-1-ene, 255
4-dodecyl phenyl sulphonate (DPS), 257, 259
DPS, *see* 4-dodecyl phenyl sulphonate
drinking water, use of ozone/hydrogen peroxide combination for micropollutant removal in, 265–277
  ozone-hydrogen peroxide combination, 266
  removal of aromatic compounds from groundwater, 267–271
    combination of ozone with hydrogen peroxide, 269
    efficiency of complete treatment line, 269–271
    ozonation, 268
    pilot plant, 267
    raw water characteristics, 267
    results, 267
  removal of geosmin and 2-MIB, 271–273
    experimental setup, 272
    by ozone/hydrogen peroxide combination, 273–275
    results, 272–273
    test solutions, 272
  removal of trichlorethylene and tetrachlorethylene from ground water, 275–276
    raw water characteristics and experimental setup, 275
    results, 275
drinking water treatment, automatic control of coagulant dose in, 157–171
  experimental methods, 158–161
  presentation of SCD, 158
  results, 161–170
    Bourbince river water, 166–170
    Seine river water, 161–166
    evaluation of SCD by batch experiments, 161–164
    on-line control experiments, 164–166
drinking water treatment, membrane filtration in, 299–310
  from feasibility to plant specifications, 300–304
    choice of membrane nature, 302
    plant specifications, 304
    validation and operating conditions, 303–304
  plant startup and operations, 304–309
    membrane manufacturing and water treatment, 304–306
    plant operation, 306–309
  ultrafiltration, 300
drinking water treatment in China, application of ozone with activated carbon for, 187–193
  cost analysis, 191
  ozonation in Chinese waterworks, 188–189
  ozone generators, 189–190
  ozone-water contactor, 190
  recent research on ozonation for drinking water treatment, 191–193
drinking water treatment plants, 207
drinking water treatment technologies, megatrends in, 325–332
  absorption, oxidation, and disinfection, 328
  automation, 332
  biological treatment, 328–329
  clarification, 329
  focus on organic matter removal, 327–328
  membrane separation, 329–332
  pressures to change, 326–327
  research orientations, 327

effective density, increase of, 11
eicos-1-ene, 255
endrin, 48

equilibrium capacity, 38
equilibrium concentration, 15
esters, 69, 251
estimation, adsorption, 77
ethanol, 74
ethers, 69, *see also* specific compound
ethyl acetate, 75
2-ethyl-1-butanol, 75
ethyl chloride, 75
ethyl ether, 74
2-ethyl-1-hexanol, 75
expert system
  additional SAR method in, 77
  for success of activated carbon adsorption, 67

fatty acids, 251, 253
FBBP, *see* fluidized biofilter bed process
FDS, *see* fluorescence decay spectroscopy
feed equipment, design requirements for, 57
Fe(III) flocculants, 99
filter acids, 300
filtration, 327
final permeability, 315
flavor profile analysis (FPA), 209, 214–216, 233, 241–245
floc break-up, 112
flocculant concentration, electrokinetic potential of flocs vs, 101
flocculants, 104–105, 300
flocculation, 105–107, 115, *see also* coagulation-flocculation
flocculation, and coagulation, 98
flocculation tests, 117
flocs, characterization and control of, 98
fluidized biofilter bed process (FBBP), 280
fluoranthene, 74, 256
fluorene, 74, 259
fluoren-9-one, 256, 259
fluorescence decay spectroscopy (FDS), 2
fluorine, 256
flux, optimization of TOC removal and, 311–324
  experimental plan/approach, 312–313
  results, 314–323
FPA, *see* flavor profile analysis
fractal aggregates, built by Al_{13} ions, 99
Freundlich equation, 19, 27
  column isotherm represented by, 24
  isotherm obtainable from column tests represented by, 22
  single-solute isotherms described by, 20
Freundlich isotherm database, compilation of, 67

fulvic acid, 290
GAC, *see* granular activated carbon
gas-liquid mass transfer, 195
GC/MS, 137, 242
gel permeation chromatography (GPC), 311, 314
geosmin, 41, 291
GPC, see gel permeation chromatography
granular activated carbon (GAC), 35, 79, 207–217
granular media depth filtration, 134

HA, see humic acid
halogenated aliphatics, 290
headloss, development of, 129
heavy metals, adsorption of, 94
hemimicelles, 1, 5
heptadecane, 255
heptadecanoic acid, 256
heptanal, 257, 260
heptan-2-one, 257
herbicides, 48
heterocoagulation, 98
heterocyclic compounds, 251
hexachlorobenzene, 74
hexachloroethane, 74
hexadecane, 255
hexadecanoic acid, 256
hexadec-1-ene, 255
hexadec-9-enoic acid, 256
9-hexadecenoic acid, 259
2,6,10,15,19,23-hexamethyl-2,6,10,14,18,22-tetracosahaene, 255
hexanal, 257
1-hexanol, 74
2-hexanone, 74
hollow fiber ultrafiltration system, 312
humic acid (HA), 81, 290
  adsorption capacity of, 82
  adsorption isotherm of, 19, 81, 87
  batch equilibria of, 32
  batch isotherm of, 25–26
  -Cr(VI), 88–90
hydrodynamic behavior, 195
hydrogen peroxide, 258
hydrolysis, 294
hydrophilic membrane, 316, 323
hydrophobic membrane, 316, 323
hydrosoluble polymers, drinking water treatment by, 115–124
  materials, 116–117
    clay, 116
    polymers, 116
    silica, 116–117
  results and discussion, 117–123

adsorption isotherm, 118–119
   flocculation behavior, 120–123
   layer thickness measurements, 119–120
   polymer-mineral surface affinity, 117–118
   techniques, 117
hydroxyhydroperoxide, 258

IAS theory, *see* ideal adsorbed solution theory
ideal adsorbed solution (IAS) theory, 19, 23, 32
initial flux, 315
initial permeability, 315
inorganic pollutants, 79
interparticle bridging, flocculation mechanism described according to, 110
iodine number, PAC characterized by, 36
ionic amphiphiles, adsorption of long chain, 4
ionic surfactants, solubility of vs temperature, 3
irreversible fouling, 316
isobutyl acetate, 75
isopropyl acetate, 75
isopropyl ether, 74
isotherm, of solute at fixed dosage of activated carbon, 30

Kamlet, linear solvation energy relationships of, 68, 71
Keggin Al_{13} structure, 104
ketones, 69, 72, 258
kinetic experiments, 266
Krafft point, 1, 7–11

lake water, ultrafiltration of, *see* flux, optimization of TOC removal and
Langmuir model, 69
Lewis-Whitman double film theory, 196
lignin, removal of, 226
lindane, 48
linear isotherm, 27
linear solvation energy relationships (LSER), 68, 70, 71, 73
liquid/solid separation, efficiency of by coagulating, 98
Log P, octanol-water partition coefficients measured as, 72
loops, formation by adsorbed flocculants, 110
LSER, *see* linear solvation energy relationships
LURE, synchrotron radiation of DCI storage ring of, 101

mass fractals, 102

maximum contaminant levels (MCLs), 50
MCLs, *see* maximum contaminant levels
membrane backflushing, 321
membrane/bioreactors, 330
membrane and depth filtration, role of colloids and dissolved organics in, 125–136
   depth filtration, colloids, and dissolved organic materials, 126–130
membrane filtration, 125, 329
   flux, fouling and, 125, 329
membrane flux, 311
membrane fouling, 133, 311
membrane molecular weight cutoff (MWCO), 314
membrane separation, 326, 327
membrane surface, 134
meta-xylenes, 256
methanol, 74
methoxychlor, 48
methyl acetate, 75
methylene chloride, 74
5-methyl-2-hexanone, 75
methylisoborneol, 41, 291
methyl nonanoate, 258
4-methyl-2-pentanone, 75
2-methyl-1-propanol, 75
2-methyl-2-propanol, 75
micropollutants, adsorption isotherms of, *see* total organics, influence of concentration change of raw water upon carbon adsorption isotherms of micropollutants and,
micropollutants, removal of, 207
mineralization, 294
model, 29
   of absorption mechanisms, 328
   derived from simulation results, 32
   influence of raw water on adsorption isotherm predicted by, 20
modified isotherm, 19, 23, 24
module conditioning solutions, 306
molar absorbance, 24
molasses number, PAC characterized by, 36
molecular connectivity, 68, 73
molecular size distribution, 143, 146, 148, 151, 155
Mont Valérian plant, 178–184
Moulle plant, 174–178
multisolute system, 20
multivariate analysis, 245
mutagens, reduction of by GAC filtration, 231
MWCO, (membrane molecular weight cutoff), 314

naphthalene, 256, 259
1-naphthol, 74
2-naphthol, 74
1-naphthylamine, 74
2-naphthylamine, 74
natural organic polymers, 289, 291, 296
natural surface waters, 157
n-butyl acetate, 75
nitrification, 291, 328
4-nitroaniline, 74
nitrobenzene, 74
nitro compounds, 267
nitrophenol, removal of, 226
4-nitrophenol, 74
nonanal, 258
nonanal acid, 258
nonanedioic acid, 257, 260
nonanoic acid, 258
nondecane, 255
non-1-ene, 255
non-4-ene, 255
nonionic surfactant, 10, 259
nonyl phenyl heptakis ethoxylate, 257, 259
n-propyl acetate, 75

octadecane, 255
octadecanoic acid, 256
octadec-9,12-dienoic acid, 256
octadec-1-ene, 255
octanoic acid, 256
octanol-water partition coefficients, 68, 72
octan-2-one, 257
odor-producing molecules, removal of, 79
odors, 291
  fishy, 239
  fruity, fragrant, and orange-like, 240, 246
  muddy, 239
  musty and earthy, 216
ofc, *see* optimum flocculation concentration
oil, removal of, 226
oligotrophic bacteria, 290
oligotrophy, 290
optimum flocculation concentration (ofc), 101, 120
organic chemicals, removal of, 281
organic membrane, 302
organic micropollutants, 2
organic pollutants, 79
ortho-xylenes, 256
oxidant, 236
9-oxodecanoic acid methyl ester, 257
9-oxononanoic acid, 257, 260
9-oxononanoic acid methyl ester, 257
9-oxononanoic methyl ester, 257

oxygen recharge, 279
ozonation, 191–193, 220
ozonation, organic removal by GAC adsorption and, 207–217
  experimental equipment and methods, 208–210
  methods, 208–210
    closed-loop stripping analysis, 208
    flavor profile analysis, 209
    optical density measurement, 208
    THM analysis, 210
  pilot plant, 208
  results and discussion, 211–216
    kinetic constants and stoichiometric coefficient, 211
    pilot plant results, 211–213
      chemical oxygen demand, 213
      total organic carbon, 213
      trihalomethanes and trihalomethane formation potential, 213–216
    CLSA-GC/MS, 214
    flavor profile analysis, 214–216
ozonation, responsibility for decrease of intensity, 247
ozonation-activated carbon adsorption, 220
ozone
  combination of with UV, 328
  disinfection by, 229
  mechanisms of action of, 236
  use in drinking water, 252
ozone, aqueous reactions of specific organic compounds with, 251–263
  discussion, 258–259
  experimental, 252–255
    extraction procedures, 253
    instrumentation, 253–255
    methodology, 252
    ozonation procedure, 253
  results, 255–258
    alkanes, 255
    alkenes, 255–256
    aromatic hydrocarbons, 256
    fatty acids, 256–257
    permethrin, 258
    surfactants, 257–258
ozone contactors, basic concepts for choice and design of, 195–206
  CT concept and hydrodynamic behavior, 204–205
  Deep U Tube, 202–204
  fundamentals of ozone mass transfer, 196–198
    absorption with chemical reaction, 197–198

INDEX

physical absorption, 196–197
hydrodynamic behavior, 198–199
mass balance of ozone in different ozone contactors, 200–202
bubbles contactors, 200–202
ozone generator, 187, 189, 192
ozone oxidation, 327
ozonide, 258, 260

p-aminobenzoic acid
adsorption equilibria of, 19
batch adsorption of, 30
correlation of background organics and, 32
isotherm of, 31
PAC, *see* powdered activated carbon
PACS, *see* polyaluminum chlorosulfate
para-xylenes, 256
particle size
distribution, 36
measurements, 312
relation between performance and, in depth filtration, 126
particle transport, 126
pathogens, use of ozone to kill, 229
Pb(II), adsorption isotherms of, 84, 85
pentadecane, 255
1-pentanol, 74
2-pentanone, 75
peracids, 258
permeate flux, 131
production of constant, 303
relation to pressure drop, 130
permethrin, 253, 258, 259
cis-permethrin, 258
trans-permethrin, 258
pesticides, 47–48, 251, 258
phenanthrene, 256
phenol, 74
adsorption isotherms of, 81, 84, 85
removal of, 226
phthalic acid, 256, 259
PIC, 112
pilot Pulsator, 158
plate counts, 291
point of zeta reversal (PZR), 12
polishing treatments, ozonation and GAC filtration as, 207
polyaluminum chlorosulfate (PACS), 99, 112
Polyani potential theory, 68, 72
polycation $Al_{13}O_4OH_{24}(H_2O)_{12}^{7+}$, 99
polyhydroxylated aromatic compounds, 256
polymeric coagulants, 329

polynuclear cationic hydroxy complexes, production of, 83
polypeptides, 290
polysaccharides, 291, 294
polysulfone, 320
polysulfone membrane, 314
potable water, pilot-plant study on advanced treatment of, 219–232
experimental process and parameters studied, 220–223
results, 223–231
Ames test, 229–231
disinfection by ozone, 229
removal of pollutants, 223–229
ammonia and nitrate, 228–229
chloroform, 226–227
COD, 223
ultraviolet absorbing compounds, 224–226
potable water treatment, evolution of tastes and odors in, 247
potential precursors, 328
powdered activated carbon (PAC), 35–65, 112, 312
characteristics, 36
cost analysis of PAC use, 57–60
factors affecting performance of, 36–38
means of applying, 39–41
improving adsorption efficiency, 40–41
reactions with treatment chemicals and calcium carbonate, 41
performance, 41–57
pesticides and herbicides, 47–49
T & O-causing compounds, 46–47
pretreatment, 323
SOCs, 50–51
THMs and THMFP, 51–57
adsorption of THM precursors, 51–54
adsorption of THMs, 55–57
prechlorination, 55, 220, 229, 231
preozonation, influence on clarification efficiency, 173–186
effect of preozonation on postozonation ozone demand, 184–185
methodology, 173–174
Mont Valérien plant case, 178–184
determination of coagulant dose, 179
determination of O_3 preozonation treatment rate, 179–182
effect of total trihaloform formation potential, 182–184
raw water, 179
Moulle plant case, 174–178
colloidal content, 175

organic matter, 177
    potential of haloforms formation, 178
    TOC, 175
    turbidity, 175–177
preozonation and algae removal, 185
pressure-driven membrane processes, 300
pressure drop, 130
primary substrate, 295
pristane, 255
privatization, 326
1-propanol, 74
2-propanol, 75
2-propen-1-ol, 75
propionaldehyde, 75
proteins, 291
p-xylene, 74
pyrene, 256, 259
pyridinium ion, 259
PZR, *see* point of zeta reversal

quality standards, tightening of, 326
quaternary ammonium carboxylic acids, 258
quaternary ammonium surfactant, 253, 257–260

radical, oxidation by, 266
rapid mix, PAC added to, 39
reaction, 195
reactors, allowing combination of oxidants, 328
regenerated cellulose membrane, 314
residual chlorine, 55
resin adsorption, 137, 138
reversible fouling, 320
rhodamine B dye, 38
Roberts-Haberer process, 40

sand filters, 329
SARs, *see* structural activity relationships
saturated aliphatics, 262
saturated fatty acids, 256, 258
saturation concentration, 6
SAXS, *see* small angle X-ray scattering, 97
scattering angle, 102
scattering vector amplitude, 102
SCD, *see* streaming current detector
SDE, *see* simultaneous distillation extraction
secondary substrates, 294, 295, 297
Seine river, water treatment, 28–32, 161–166
separation, combination of with reaction in recirculating loop, 330
series resistance model, 320
silica, 116

colloids, 99
    -liquid interface, fractal aggregates adsorbed in flat conformation at, 110
    sol, rapid-mixing of, 99
simulation
    based on IAS theory, 20
    of batch adsorption equilibria, 22
    execution for water containing solutes, 19
    for imaginary raw water, 21–24
simultaneous distillation extraction (SDE), 242, 246
single-solute isotherms, 19
single-solute system, 20, 85, 87
sludge dewatering, 327
small angle X-ray scattering (SAXS), 97, 101
SOCs, *see* synthetic organic chemicals
solid-liquid interface, adsorption thermodynamics at, 1–17
    adsorption of surfactants, 4–7
    application to removal of some organic reagents in water treatment, 11–15
        removal of diethylphtalate and dibutylphtalate, 14–15
        removal of long chain sodium soaps during flocculation with aluminum hydroxide gels, 11–14
    how surfactants adsorb on hydrophilic surface, 7–11
        isotherms preformed at temperature below Krafft point, 11
        isotherms preformed at temperature higher than Krafft point, 7–11
    physicochemical properties of surfactants in solution, 3–4
solid-phase concentration, ratio of liquid-phase concentration to, 19
solids contact slurry recirculating clarifier, 40
solvatochromic parameters, 71
spreading pressure, 19, 20
squalane, 255
static membrane adsorption, 311, 318, 319
statistical analysis, 247
Stokes' law, 107
streaming current detector (SCD), 157, 161
structural activity relationships (SARs), 67
surface water treatment, 311, 312
surfactant adsorption, mechanisms of, 2
surfactants, 2, 251, 257–258
surrogate parameters, behavior of total organics represented by, 20
synchrotron radiation, 101
synthetic organic chemicals (SOCs), 35, 289, 296

INDEX

synthetic organic compounds, 50

taste and odor problems, ozonation of organic compounds causing, 233–250
  Lyonnaise des Eaux, Morsang Water Treatment Plant, 237–247
    conventional treatment of taste and odors, 237–239
    correlation between sensory analysis and chromatographic profiles, 245–247
    correlation of flavor profile and chemical analyses, 241–245
    ozonation treatment, 239–241
    statistical methods, 245
  water treatment for removal of taste and odor compounds, 235–237
taste- and odor-causing (T & O) compounds, 35, 46–47
tastes, 291
  astringent, 240
  fishy, 239
  fruit, orange-like, and fragrant, 246
  muddy, 239
  musty and earthy, 216
  plastic and pharmaceutical, 240
  removal of, 207
TCE, *see* trichloroethylene
1,1,2,2-tetrachloroethane, 74
tetrachloroethylene, 74, 275–276
tetradecane, 255
tetradecanoic acid, 256
tetradec-1-ene, 255
THM, *see* trihalomethane
THMFP, *see* trihalomethane formation potential
threshold odor number (TON), 46, 234, 267
TOC, *see* total organic carbon
TOC/E260, 143, 147, 153
T & O compounds, *see* taste- and odor-causing compounds
T & O control, 46
toluene, 50, 256
TON, *see* threshold odor number
total organic carbon (TOC), 51, 80, 174, 304, 311
total organics, influence of concentration change of raw water upon carbon adsorption isotherms of micro-pollutants and, 19–34
  effect of dilution or concentration of raw water on adsorption isotherm of total solutes, 21–26
    batch isoform of commercial humic acid, 25–26
    simulation for imaginary raw water, 21–24
    IAS theory, 20–21
  influence of background organics upon the adsorption equilibrium of certain components, 26–32
    batch isotherm of p-aminobenzoic acid in Seine River water, 28–32
    simulation for imaginary raw water, 27–29
total organics, isotherms of, 32
total solid-phase concentration, 19, 26
  adsorption equilibrium dependent upon, 27
  correlation of spreading pressure and, 28
total solutes, isotherm of, 22
total trihaloform formation potential (TTFP), 182
toxic pollutants, 70
2,4,5-TP, 48
Traube's law, 4
treatability evaluation, by simple and rapid method, 143–155
  experimentals, 144–145
    coagulation and biological treatment experiments, 144–145
    operational condition for HPSEC, 144
    sample preparation, 144
  results and discussion, 146–154
    biological treatment experiments, 152–155
    characteristics of apparent molecular size distribution of water and wastewater, 146–148
    coagulation experiments, 148–152
    column standardization and TOC recovery, 146
treated and raw water, preliminary analysis of organic contents in Guangzhou City, 137–141
  collection and preliminary treatment of water samples, 137–138
  preliminary evaluation of removal of organic chemicals by treatment facilities, 141
  qualitative and quantitative analysis of organic contaminants, 138–139
  results, 139–140
1,2,4-trichlorobenzene, 74
1,1,1-trichloroethane, 74
1,1,2-trichloroethane, 74
trichloroethylene (TCE), 50, 74, 275–276

2,4,6-trichlorophenol, 74
tridecane, 255
tridec-1-ene, 255
trihalomethane (THM), 35, 52
  adsorption of, 55–57
  formation of during drinking water treatment, 51
  THMFP and, 51–57
trihalomethane formation potential (THMFP), 51, 303, 311, 316
trimethyl alkyl ammonium chloride, 258
1,2,3-trioxolane, 258
TTFP, see total trihaloform formation potential
turbidity, 174, 175, 291, 311
  removal, 329
  spikes, 304

ultrafiltration, 299, 311, 315
ultraviolet (UV)-absorbing materials, 311
undecane, 255
undec-1-ene, 255
unsaturated aliphatics, 262
unsaturated fatty acids, 256, 259

U.S. Environmental Protection Agency (U.S. EPA), 50, 139
UV-absorbance, 24, 174, 185, 302
  batch equilibria of humic acid measured by, 32
  isotherm measured by, 19, 25, 31
  modified isotherms based on, 25–26

valeraldehyde, 75
vinyl acetate, 75
viruses, use of ozone to kill, 229
VOCs, see volatile organic chemicals
volatile organic chemicals (VOCs), 36, 50

water
  biologically stable, 289
  organoleptic properties of, 234
  quality, 223
  removal of particles from, 134
  resources, reduction in, 326
  soluble polymer, 115
wavelength, 102

XAD-2, as adsorbent, 137